Service Management

SERVICE MANAGEMENT

A Harvard Business Review Paperback

Harvard Business Review paperback No. 90090

ISBN 0-87584-298-4

The *Harvard Business Review* articles in this collection are available as individual reprints. Discounts apply to quantity purchases. For information and ordering contact Operations Department, Harvard Business School Publishing Division, Boston, MA 02163. Telephone: (617) 495-6192, 9 a.m. to 5 p.m. Eastern Time, Monday through Friday. Fax: (617) 495-6985, 24 hours a day.

Editor's Note: Some articles in this book may have been written before authors and editors began to take into consideration the role of women in management. We hope the archaic usage representing all managers as male does not detract from the usefulness of the collection.

Printed in the United States of America by Harvard University, Office of the University Publisher.
93 92 91 5 4 3 2 1

Contents

The Service Industry

Service Companies: Focus or Falter
William H. Davidow and Bro Uttal
3
Service companies must target and supply the needs of a particular market or risk losing business.

Lessons in the Service Sector
James L. Heskett
12
To compete successfully in the service sector, operations and management must be synchronized.

Will Services Follow Manufacturing into Decline?
James Brian Quinn and Christopher E. Gagnon
21
The U.S. service sector may be headed toward the same disastrous global noncompetitiveness that plagues manufacturing.

A Radical Prescription for Hospitals
Jeff Goldsmith
31
Soon, hospitals will bring services to the patients as well as patients to the hospital.

Strategies for Customer Relations

Zero Defections: Quality Comes to Services
Frederick F. Reichheld and W. Earl Sasser, Jr.
41
To learn how to keep customers, track the ones you lose.

The Profitable Art of Service Recovery
Christopher W.L. Hart, James L. Heskett, and W. Earl Sasser, Jr.
49
How the best companies turn complaining customers into loyal ones.

The Power of Unconditional Service Guarantees
Christopher W.L. Hart
59
A service guarantee can ensure satisfied customers *and* become a powerful management tool.

My Employees Are My Service Guarantee
Timothy W. Firnstahl
69
Give your employees the authority to honor your service guarantee, and you will satisfy *them* as well as your customers.

Adding Value to Products

Beyond Products: Services-Based Strategy
James Brian Quinn, Thomas L. Doorley, and Penny C. Paquette
77
U.S. industry can regain strategic advantage by developing key service capabilities.

The Service Factory
Richard B. Chase and David A. Garvin
85
In the future, factories will become service centers, where managers and workers respond directly to customer needs.

Exploit Your Product's Service Life Cycle
George W. Potts
95
The service life cycle can prove profitable if given the same shrewd attention as the product life cycle.

Competing on the Eight Dimensions of Quality
David A. Garvin
99
High quality means pleasing customers, not just protecting them from annoyances.

Case Studies

The Case of the Complaining Customer
Dan Finkelman and Tony Goland
111
How much service does a company or store owe a customer?

The Case of the Not-So-Supermarket
Janice McCormick
121
The service dilemma: find enough workers, upgrade their skills, and improve productivity—without giving away the store.

The Service Industry

Without a strategy, you don't know who your customers are.

Service Companies: Focus or Falter

by William H. Davidow and Bro Uttal

In an industry bedeviled by rising costs, Shouldice Hospital near Toronto, Canada is a model of cost-bashing productivity. Surgery patients stay at Shouldice for 3½ days compared with 5 to 8 days at most hospitals. Its doctors perform many more operations a year than elsewhere, yet they are paid less than they would make in private practice; nurses attend to several times as many patients as they would in most other hospitals. Patients care for themselves; they get to the operating room on their own steam, walk to the recovery room, and take their meals in a common dining room.

Does this sound like a low-quality, uncaring production line? It's just the opposite. Measured by how often patients need repeated treatment for the same problem, Shouldice is ten times more effective than other hospitals. Ex-patients are so delighted with the Shouldice experience that they hold an annual reunion to commemorate it. The January 1988 jamboree at the Royal York Hotel in Toronto attracted some 1,500 "alumni."

There's something else unusual about this picture of great service: most sick people couldn't get admitted to Shouldice. If you have a broken bone, gallstones, or clogged arteries, apply elsewhere. Shouldice accepts only one type of patient–people whose single complaint is a hernia. Even then, the hospital will reject hernia victims who have histories of heart trouble or who have undergone surgery in the last 12 months or who weigh more than Shouldice recommends.

A tightly focused service strategy is the key to Shouldice's performance. By segmenting the market of sick people according to their complaints, then concentrating on one segment that is inexpensive to serve, Shouldice has optimized its operations, fulfilled its mission–and enjoyed a handsome return.

After doing hundreds of hernia repairs a year using a special technique, Shouldice's doctors have become highly proficient and productive. The hospital usually can avoid using general anesthesia because local anesthesia, which is safer and cheaper, works well for hernia repair. Patients recover from surgery faster if they're up and about, so Shouldice avoids paying for a fleet of wheelchairs and gurneys, armies of aides to push them, and banks of wide elevators. Instead, it gives patients hallways with comfortable carpeting, staircases with low risers, and acres of well-groomed grounds on which to stroll. The central location of television sets and toilets encourages patients to walk around even as it saves the hospital money.

William H. Davidow is a general partner in a California high-technology venture capital firm, Mohr, Davidow Ventures. Bro Uttal is a management consultant.

Concentration on hernias enables Shouldice to produce a highly competitive core service. But that's not the main reason for Shouldice's success. Other organizations, notably the Lichtenstein Hernia Institute of Los Angeles, do low-cost repairs that work at least as well. Consumers, most of whom have medical insurance, are somewhat insensitive to health care costs. And those who have personal physicians tend to follow their advice, which normally is to get a hernia fixed at a hometown hospital.

At Shouldice Hospital, patients make their own way to the operating room.

No, people flock to Shouldice mainly because they hear from former patients that being there is a great experience. The hospital earns this enthusiastic word of mouth because it adds significant (though intangible) value to its core service. Most people can't accurately judge the quality of a hospital's core service. What they can and do appraise, at least when they're not very sick, are experiences like the welcome received at check-in, the behavior of fellow patients, and the attentiveness and perceived competence of doctors and nurses.

> People Express fatally blurred its market (backpack types) when it went after business travelers.

Could Shouldice produce outstanding customer service absent its clear strategy? Probably not without completely redesigning its physical and social systems and sending its rates sky high. Shouldice is ill-suited for treating people with broken legs or weak hearts, those who can't walk much or who have had major surgery and need intravenous feeding, or patients recovering from plastic surgery (they seldom want to socialize). Shouldice managers have thought about undertaking minor eye surgery and repairing varicose veins and hemorrhoids – procedures fairly similar to hernia repair in their demands on facilities. But "fairly similar" isn't close enough. Shouldice has decided to stick with the segment it knows best and serves most effectively.[1]

In contrast, fuzzy or conflicting strategies make good customer service impossible. Look at People Express. In 1981, it focused tightly on budget travelers – students, backpackers, vacationers, and others willing to sacrifice convenient schedules and airport gates for low fares. Customers actually enjoyed the airline's widely advertised "no frills" service, which gave them the options of bringing their own food or paying extra for a snack in flight, of carrying their luggage or paying $3 a bag to check it, and of buying a ticket on board instead of in advance. In five years, People Express's fleet grew from 3 airplanes to 117, and its revenues soared from $38 million to nearly $1 billion.

But People had overexpanded. Every day, the airline had thousands of seats available, but its steady customers wanted to fly mainly on weekends and during the summer and other vacation times. The glut of empty seats produced a $3.7 million loss in 1984. So People scrambled to get more revenue from its fleet. It scheduled each plane for as many flights as possible and it grossly overbooked to compensate for customers who made reservations but didn't show up. The airline also went after business travelers, whose demand is heaviest Monday through Friday and nicely complements the weekend demand pattern of budget fliers. To attract executives, the airline pitched its luxurious first-class service, complete with leather seats.

Nearly everything People Express had done to serve the budget-minded conflicted with this new strategy. Businesspeople intensely dislike inconvenient schedules and gates, paying extra for checked baggage and for meals, and waiting until the last minute to buy tickets. People's tight scheduling meant that many flights were late, and overbooking meant that it often turned away customers who had reserved seats. The carrier earned a new name, "People Distress."

1. See James L. Heskett, *Managing in the Service Economy* (Boston: Harvard Business School Press, 1986).

Business travelers stayed away, and the airline racked up a $300 million net loss in 1986, when Texas Air took it over. If it had continued to focus tightly on budget fliers and had sold off or leased planes to cut capacity, People Express might still be around. But it broadened, and irrecoverably blurred, its original winning strategy.

Markets and Market Segments

To some managers, developing a strategy for customer service may sound like a waste of time. How much strategy do you need to send out a repairperson or to adjust an erroneous bill? Yet even those seemingly simple activities won't do much for customer satisfaction or corporate profits unless they are part of a considered strategy. Without one, you don't know who your customers are, how much they value different aspects of service, how much you must spend to satisfy them, and how big the payoffs are likely to be. Without a strategy, you can't develop a concept of service to rally employees or catch conflicts between corporate strategy and customer service or come up with ways to measure service performance and perceived quality. In short, without a strategy you can't get to first base.

Developing a service strategy is an essential step toward choosing an optimal mix and level of service for different customer sets. Provide too little service, or the wrong kind, and customers will leave; provide too much, even the right kind, and your company will go broke or price itself out of the market.

Proof of that truism comes from a General Electric experiment that varied the levels of repair service for out-of-warranty appliances over a two-year period. GE discovered that repair service is very sensitive to the law of diminishing returns. At a certain point, each incremental investment in service starts to yield lower returns than the previous investment. The only way to find that point is to segment customers, find out how much they value different levels of service, and estimate the costs and benefits of serving them well.

Simply charging ahead with extraordinary service doesn't guarantee a high return, as Service Supply Corporation of Indianapolis, the "house of a million screws," should know. If Service Supply had a cogent customer service strategy, it would stratify its 15,000 customers according to their sensitivity to stockouts of different classes of fasteners, then adjust its inventory levels and prices to address those tiers with the highest potential returns. As it is, Service Supply gives outstanding service to *all* its customers, regardless of what service they need or their willingness to pay, and it can't charge enough on average to bring profits up to industry norms.

The essence of any customer service strategy is to segment the customers to be served. As with classic market segmentation, the goal is to isolate a reasonably homogeneous set of customers that can be served at a profit. But customer service segments differ from the usual market segments in significant ways. For one thing, they tend to be narrower. While many different kinds of customers may be happy to buy the same product, they're less likely to feel that their expectations have been met or exceeded if the service they receive is standardized and routinized. Service expectations, after all, are highly personal.

Moreover, marketing segmentation focuses on what people and organizations *need*, while customer service segmentation focuses on what they *expect*. Marketers tend to use immediate sales to judge whether a segment is vaiid. Since purchasing decisions look binary – the sale is made or it isn't – the validity of marketing segments seems easy to assess.

During the week, People Express had a huge surplus of empty seats.

Occasionally, marketing segments and customer service segments are the same, especially for service companies that have well-focused, comprehensive marketing strategies. Those strategies can meet a broad range of service expectations. Federal Express's strategy, for example, is to meet expectations for all the actions and reactions that customers perceive they have purchased, including not only pickup and delivery but also documentation and information about shipments.

More often, though, defining customer service segments means rethinking overly broad market segments and the ways in which they impede superior service. Intelligent segmentation can transform the productivity and profitability of customer service operations, as Shouldice's strategy shows.

It's hard to get big productivity gains by substituting capital and technology for labor, since high-touch customer service by definition means lots of flexible, warm, human contact. So managers often seek to substitute low-touch service for the high-touch kind – offering, for instance, automatic checkout from the hotel instead of the usual front-desk checkout.

> By segmenting your market, you can better match supply and demand.

The problem is that some customers welcome low-touch substitutes while others view them as cost-cutting measures that seriously reduce the quality of service. Compared with business travelers, tourists staying at a grand hotel have a far less favorable view of automatic checkout. They may see it as a jarring, impersonal note in their hotel experience. Veteran users of large home appliances may prefer getting repair instructions over the telephone, while novices expect a human repair expert to show up. Only after a company has segmented its customers and chosen which ones to serve can it figure out where to substitute low touch for high, thus improving productivity without imperiling customer satisfaction.

Segmentation also is key to solving one of the thorniest problems in customer service – matching supply and demand. Manufacturers with overcapacity will produce for inventory or shut down machines and lay off workers. When demand exceeds supply, they draw down inventory, run their factories continuously, hire extra shifts, and call on the capacity of other suppliers.

Service providers have fewer options. Service can't be stored in inventory. Seldom can you substitute another company's service for your own and keep meeting customer expectations, since those expectations refer to a unique experience, not an interchangeable product. A sudden addition to or reduction in capacity (people, mainly) ensures shoddy quality, or even a breakdown of the service "factory," as anyone can attest who has stayed at a large hotel when it's hosting a convention. Since demand for customer service isn't homogeneous – every customer demands somewhat different versions of excellent service – service organizations need idle capacity for adapting their product to mixed demand. By some estimates, service quality drops off sharply when demand exceeds as little as 75% of theoretical capacity.[2]

Until it segments, a hotel or a field force of computer engineers may see huge and apparently random fluctuations in demand. But segmenting usually shows that the overall pattern is made up of several smaller, more predictable, and therefore more manageable patterns. For instance, convention visitors, ordinary business travelers, foreign tourists, and vacationing families all contribute to variations in demand for hotel service. The distinctive patterns of their demands can be forecast.

Segmenting may even reveal that some kinds of demand are undesirable and should be finessed, like the demand on a bank to make change for noncustomers and the demand on a fire department to rescue cats from trees.[3] But simply ignoring demand is nearly always dangerous. Disgruntled noncustomers, like the grandmother whose cat is up a tree, can spread as much bad word about a company as genuine customers can. Better than ignoring inappropriate demand is lowering the expectations of noncustomers, say by broadcasting to cat owners how busy the fire department is fighting fires or by creating disincentives like a $20 charge for restoring arboreal cats to earth.

Among the most powerful ways to vary service capacity with demand is to expand the role of customers in producing service, in effect making them coproducers. To some degree, the customer always helps produce service, by reading a manual, returning an appliance for repair, filling out an air bill, or participating in the service ritual at a fancy restaurant. Finding opportunities to expand the customer's role and move toward self-service often depends on savvy segmentation.

Automated teller machines, for example, increase the role of customers in producing some banking services and thus represent highly flexible capacity. The

2. See James L. Heskett, *Managing in the Service Economy* (Boston: Harvard Business School Press, 1986), p.38.

3. The cats come courtesy of Christopher H. Lovelock, *Services Marketing* (Englewood Cliffs, N.J.: Prentice-Hall, 1984).

greater the demand on the machines, the greater the capacity customers provide (at least until the waiting line gets too long). But targeting the right kinds of customers for ATMs is crucial. Retirees and other older people, expecially in smaller towns, tend to resist the machines. So do many of the very wealthy, who expect the attention of human beings in return for their lucrative business.

For some, getting repair instructions over the phone is sufficient.

Carving out segments is a matter of life and death for some companies. A good example is banking, whose core services are close to being pure commodities. Pundits have forecast the death of the small-town bank ever since the early 1980s, when deregulation started opening local markets to more efficient regional and money-center banks. Many local banks have in fact closed their doors or disappeared into the maws of acquisition-hungry superregionals.

Some local banks, though, have prospered despite the competition by focusing on certain customer segments whose service expectations the big boys don't satisfy. University National Bank & Trust of Palo Alto, California has just one office. The bank discourages small depositors by setting a high monthly fee for checking accounts, $20, and waiving it only for average monthly balances of at least $3,000. Deadbeats need not apply: the bank accepts new customers only with a referral or after a credit check. Bounce two checks and UNBT will close your account.

Yet UNBT keeps gaining market share and consistently earns a return on assets more than twice the average for California banks. Like its mammoth competitors Wells Fargo, First Interstate, and Bank of America, UNBT has targeted individuals with high net worth, generally more than $500,000. Unlike its competitors, UNBT serves *only* those customers, and it serves them with an array of services that includes free traveler's checks, cashier's checks, and stop-payment orders, streamlined approval of large loans, free shoe shines on visits to the bank, and a sack of sweet onions gratis every July. UNBT successfully targeted the people in the community who wanted lots of hands-on service.

More important than any of these features, however, says CEO Carl J. Schmitt, "was the concept of keeping the 1/2 of 1% of customers who are bad guys out of the bank's customer base. It is that 1/2 of 1% that causes the large banks to create the arbitrary rules that in turn create the hassle experienced in the large bank."

The "two bad checks and you're out" policy improves the bank's service in an interesting way. For one thing, local merchants know all UNBT checks are good, so they honor them as if they were cash. For another, tellers don't have to spend time checking the customer's credit. In almost all cases, the bank cashes customer checks without inquiring whether they have money in their accounts. This reduces lines at the teller windows. So the bad-checks policy leads to better credit outside and faster service inside the bank.

If you can carve out a niche by offering platinum service, can you also carve out a niche by offering virtually no service? Sure, if you spot the niche and handle it right. One market segment that is too "cheap" to pay for service is the college population. The East Lansing State Bank has prospered mightily by taking on Michigan State students with no-minimum-balance, all-ATM checking accounts. The students get five checks a month for a monthly service charge of $1, and they pay 50 cents for each window transaction. The bank knows where it wants the students—and it isn't at the teller windows clogging up service to its regular customers.

Classifying Customers

In reality, most companies can't segment their customers as sharply as Shouldice Hospital and University National Bank & Trust do; the set of existing customers is usually too diverse to be stuffed into a single pigeonhole. The problem is most difficult, of course, for manufacturers because they usually sell to a wide range of buyers and use distribution channels that insulate them from direct knowledge about their customers.

Nonetheless, any company can carve out useful segments, rank their attractiveness, and develop a focused service strategy by examining a few key characteristics of its customers and its business. The traits to look at first are obvious financial ones. How do the typical size of a sale and the likelihood of repeat sales vary among customers? What are the costs of giving superior service to different types of customers? Approximating the answers produces a rough segmentation and a working notion of the benefits and costs of different service strategies.

In any well-run business, service levels normally correlate closely with size of sale. For a $150 suit off the rack at a discount store, you won't get free alterations, let alone any advice from a salesperson about fashion trends, the fit of the suit, or appropriate accessories. But in keeping your appointment with Bijan, which sells $1,500 suits behind barred storefronts on Rodeo Drive in Beverly Hills and on Fifth Avenue in New York, you can expect head-to-toe fashion advice, any alteration you care to name, and a phone call after the sale to make sure that everything fits as well as you thought it would.

Note that people who buy discount suits don't necessarily get worse service. They do get less of it, but whether they perceive that consequence as low-quality service depends on their expectations.

When automotive analyst Martin Stein examined 26 auto service programs and their impact on customer satisfaction, he was not surprised to find that the more expensive the car, the more elaborate the service. The manufacturers of Mercedes and Acura take the "open pocketbook" approach: whatever customers want, they get, including 24-hour roadside assistance.

The companies with the worst service quality and the lowest customer satisfaction weren't those making the lowest priced cars or offering the lowest absolute levels of service, but those lacking a service strategy and failing to match service levels with the size of sale. Except in the case of Cadillac, General Motors took a cosmetic approach: it sent service managers through a training program where they learned to refuse requests for service politely, and it used "800" numbers to field complaints. "This approach backfires," according to Stein. "Customers say that General Motors is 'supercilious' and that 'they are overly courteous but don't want to fix your car.' Polite evasion antagonizes people more than if you just say you can't do it. Tactically, there's nothing wrong with "800" numbers, but given GM's lack of a service strategy, its managers don't see any positive change in customer satisfaction."

When their product permits it, sophisticated companies look for customer segments that are less costly to serve. Part of Shouldice's secret is that basically well people cost less to serve than the seriously ill. Older, more conservative car owners who live in the suburbs are less costly to serve than people in their twenties who live in urban and rural areas, a fact that canny auto dealers and insurance companies appreciate. Companies with strong central-purchasing offices tend to be less expensive to serve than those that let each branch or division make its own buying decisions.

> You want to keep your best customers, so rank them by their value.

Besides traits like health, age, and geographical location, some segments have other characteristics that cut the costs of serving them. Many customers are happy to share the burdens of service, like shoppers at self-service stores and industrial buyers who order parts and supplies through computers linked directly to their suppliers. Other customers are willing to adjust their demand for service to accommodate the supplier, like the budget-minded who accepted People Express's no-frills tack, like Shouldice patients who shave their own groins and abdomens before surgery, and like utility customers who limit their electrical usage during peak demand hours.

The cost of service also is affected by customers' knowledgeability and by their ability and willingness to cooperate in getting service. Buyers of semiconductor chips expect the products to meet their specifications, and living up to specs is a crucial aspect of the service that chip makers offer. But unless the buyer adjusts its automated test equipment to perform the same way the chip maker's does, the buyer will reject batch after batch of chips for being out of spec, even though, according to the chip maker's test equipment, the parts were fine when they left the factory.

Competent customers are what Network Equipment Technologies seeks for its communications multiplexers. The California manufacturer focuses on large companies with sophisticated communication needs and able communication management teams. Its equipment is designed with diagnostic features that make it easy for NET to diagnose failures remotely. With telephone support, most customers can fix the equipment themselves. As a result, service engineers stay at home; the company needs to dispatch them on a customer call in only about 240 of every 1,000 incidents. NET can extend service to competent customers cheaply, and on top of that, customers get their equipment up and working again faster. That makes for happier customers.

Classification of customers by the value they place on service and by their service expectations often generates a rough idea of the cost of satisfying them. But which customers are the best targets? Those who are the most valuable compared with the likely costs of serving them. While, as we suggested, size and frequency of purchase are good indicators of value, there are many others, such as: customers whose demand is likely to grow faster than average; influential customers, who will generate powerful word of mouth; and loyal customers. Buyers who are especially demanding, sophisticated, or technologically advanced often go to the top of the list because serving them gives a supplier insight into the needs of more ordinary customers.

Ranking customers by their value is essential for any service operation that must live with big swings in aggregate demand and can't adjust capacity quickly to meet those swings. Ranks or tiers are the key to allocation. When capacity is short, smart suppliers give top-tier customers first claim on service and cut back on service to lower value customers. That's what popular restaurants do at peak dining times by finding tables for regulars and keeping newcomers waiting, and it's what American and Japanese semiconductor makers do when capacity is short and they fill the orders of long-term customers first.

Without assigning its customers to tiers, a company that serves numbers of segments has difficulty getting the most out of its service capacity. Stretching and straining to satisfy every segment, it may end up giving low-quality service to all customers, not just to the less desirable ones in lower tiers.

Service companies have discovered through painful experience that using the same organization to serve different market segments, or to provide very different services to the same segment, seldom works. Frito-Lay, the PepsiCo subsidiary that leads the potato chip business, is famous for its customer service. A corps of 10,000 route drivers visits most stores, many of them small grocers, two or three times weekly to ensure that the stock is fresh—and prominently displayed. At one point, the company decided to add packaged cookies that they would deliver in the same way and at the same time.

The attempt failed. Frito-Lay's service infrastructure, very efficient for potato chips, was too specialized to handle cookies well. As Leo Kiely, senior vice president for sales and marketing, explains it: "On the surface it looked easy, but there were underlying problems. Our other products turn over in about seven days, which is very rapid. Cookies have a much slower turn, with a 60- to 90-day shelf life. That's a different inventory problem for our drivers, so our regular visit two to three times a week proved inefficient for cookies."

While handling many customer segments piles on complexities, few companies can afford the luxury of addressing just one segment. The challenge of exploiting service capacity forces them to diversify at least a little, as airlines do in targeting both business and vacation travelers. Moreover, the same customer is often a different type of customer at different times. Business executives go on vacation. New bank customers, who value location and convenience most, quickly become established bank customers and start giving the highest rankings to different criteria, like operational accuracy, dependability, and friendliness.

The best practice is to pick segments that are as similar as possible and to keep your priorities straight. People Express failed to do that.

What Do Customers Expect?

After segmenting your market so that you can target your customers, the next step is to find out what they want and expect. This takes research and analysis.

The payoffs for good analysis are tangible sales and profits. In Norway, where customers are particularly sensitive to service, Toyota used to think of cars as simply products. Then the company began looking into the total car experience and asking what kinds of service customers expected. Discovering that they were concerned not just about reliability and performance but also about the difficulties of buying cars and car insurance and about the anxieties of repair, Toyota used its large customer base to bargain for very competitive financing and insurance services, and it started offering free diagnostic service. In 1985 and 1986, Toyota's sales in Norway rose more than 30% and its earnings went from the equivalent of $12 million to $22 million.

It's tempting to forgo analysis because you assume you know what customers expect. But assumptions don't make effective customer service strategies. Inward-looking companies that are guided by industry norms and their own past practices end up with inappropriate strategies, lower market shares, and anemic profits. Time after time, studies have shown large differences between the ways that customers define service and rank the importance of different service activities and the ways that suppliers do.[4]

Numbers of companies have triumphed by filling the gap between what customers see as good service and what competitors think it is. Tom Ford, founder and head of the Ford Land Company, perceived that most developers of office buildings see their businesses as financial plays where return on assets is paramount. Most commercial tenants, he noted, are convinced that their landlords are trying to squeeze the last penny of return from their properties.

One key to success is to fill the gap between what customers see as good service and what competitors think it is.

So Ford decided to stress tenant service in the 34 office and industrial buildings he put up and operates on the peninsula south of San Francisco. He doesn't charge for extras like installing electrical outlets, laying computer cable, or changing names on doors and building directories. When remodeling creates a racket that disturbs neighboring tenants, Ford brings the neighbors wine and flowers to show he's at least aware of the problem.

The result? Since 1977, Ford's properties have been 100% leased, a remarkable record in an area where vacancy rates have been running between 10% and 30%. A competitor who opened an office building next to Ford's flagship property in Menlo Park tried to draw away Ford's tenants by offering months of free or reduced rent. None of the tenants would bite, and the new building was only 50% occupied 15 months after opening.

Good service has nothing to do with what the provider believes it is; it has to do only with what the customer believes is true. Good service results when the provider meets or exceeds the customer's expectations. Do less than the customer expects and the service is bad. Do what is expected and the service is good. Exceed by a great amount what is anticipated, as Tom Ford does, and the service will be superior. This is why the providers of good service have to be extremely careful to set the customer's expectations at the proper level.

Levels of expectation are why two organizations in the same business can offer far different levels of service and still keep customers happy. It is why McDonald's can extend excellent industrialized service with few employees per customer and why an expensive restaurant with many tuxedoed waiters may be unable to do as well from the customer's point of view. It is why diners like the folksy service at New York City's upscale steak house, Christ Cella, and would perhaps be disappointed with the same type of service at a fancy French restaurant.

Setting customer expectations at the right level can be a very tough job even for service experts. Consider Nordstrom, the most service-oriented large fashion retailer in the United States. Nordstrom's reputation for good service has gotten out of control. Stories about heroic service deeds of Nordstrom employees, fueled by the word of mouth spread by enthusiastic customers, are threatening to ruin Nordstrom's service. Why? Because customers enter the store with unrealistically high expectations.

In an effort to rein in expectations, Nordstrom has stopped talking about service to the press. Too much publicity about a company's good service can be bad. Likewise, fire departments should stress their expertise in fighting fires, and keep mum about rescuing animals, if they want to lower the expectations of owners of tree-climbing cats.

There are severe limits to any company's ability to set expectations. The strictest one is reality. Few patrons of a high-priced hotel can be led to expect anything other than luxury service, and few people who have experienced bad service can be persuaded to expect anything else. As Martin Stein, the auto-industry analyst observes, "General Motors's advertising campaign for Mr. Goodwrench, the GM dealer's mythical service expert, doesn't work because people doubt that the quality of service being

When remodeling work disturbs tenants, Tom Ford's company placates them.

4. See, for example, Norman E. Marr, "Do Managers Really Know What Service Their Customers Require?" *International Journal of Physical Distribution and Materials Management*, vol. 10, no. 7, 1980, p. 433.

advertised will be available. They may be looking for Mr. Goodwrench, but they aren't finding him." Trying to set expectations that vary widely from the realities customers perceive is futile.

Expectations are formed by many uncontrollable factors, from the experience of customers with other companies and their advertising to a customer's psychological state at the time of service delivery.

> If customers' expectations about your service get too high, you may be in trouble.

Strictly speaking, what customers expect is as diverse as their education, values, and experience. The same advertisement that shouts "personal service" to one person tells another that the advertiser has promised more than it possibly can deliver.

Yet some straightforward tactics can help bring expectations into line with the service strategy. The job is basically the same as positioning a company or product in the marketplace. Service positioning starts with four givens: the segments targeted, the expectations of those segments, the strategy for exceeding those expectations, and the positions of competitors, that is, the images they have created for their companies in customers' minds.

A winning service position meets two criteria. It uniquely distinguishes a company from the competition, and it leads customers to expect slightly less service than a company can deliver. That's what Avis did years ago by positioning itself as the second-place car rental company that has to try harder, and it's what Avis is doing today by portraying itself as the rent-a-car outfit that tries harder because the employees own the company. Maytag has done the same, positioning its washing machines as so reliable that the Maytag repairman is bored to death.

The tools used to position customer service operations are the same communication tools any marketer uses – advertising, promotion, public relations, and everything else that affects the all-important word of mouth. But sending messages about service is different from most forms of marketing communication. Since service is intangible, advertising has a special mission to dramatize service in ways that make the benefits clear and real.

All forms of communication should be tightly focused on the target segment because customers' expectations about service are affected strongly by the other kinds of customers they see. Reaching the wrong segments can be a disaster. A business traveler checking into a budget motel revises her expectations radically if she sees a drunk asleep in the lobby.

Positioning customer service differs from normal positioning in other ways as well. Customers are hypersensitive to tangible service clues like uniforms, repair trucks, brochures, and hotel lobbies. Often they can't tell that a service has been performed without some additional evidence like the elaborate receipts car mechanics make out or the strip of paper hotels wrap around toilet seats to let guests know the toilet has been cleaned. Customers' expectations of service rise and fall markedly because of seemingly minor clues or tip-offs like these.

The key to successful positioning of customer service is not to create expectations greater than the service your company can deliver. The whole organization must be together on this. Network Equipment Technology, for one, disciplines salespeople who overpromise. NET realizes that keeping expectations at just the right level – slightly below perceived performance – is a constant challenge.

Three Steps to Take

Great service providers inform customers about what to expect and then exceed the promise. Not all customers want or deserve high levels of service, but they are entitled to what they have been promised, explicitly or implicitly.

The strategy of great service providers has a rather familiar ring. Almost all of these companies have taken the same three steps:

First, they segmented the market carefully and designed core products and core services to meet the needs of the customer base. They realized that not all customers who bought the same product or service had the same service needs.

Second, they realized that only the customer knew what he or she wanted. Therefore, they researched the needs of the customer base thoroughly, both with formal programs and by paying close attention to what the customer was saying.

Third, they were careful to set the customer's expectations at the right level. They underpromised and overdelivered.

Providing good service is a towering challenge. There are many reasons businesses and professions fail in the endeavor, even when they have a perfect strategy. Without a focused service strategy, however, meeting the challenge becomes impossible.

Reprint 89403

James L. Heskett

Lessons in the service sector

☐ A large food and lodging company creates and staffs more general management jobs than any ten manufacturers of comparable size. This company, like many others dispensing high customer-contact services, has eliminated functional lines of responsibility between operations and marketing. In its planning the company routinely combines operations and marketing with what I call a strategic service vision.

☐ The most profitable large American company assumes daily the task of managing a work force of window washers, cooks, and maintenance personnel. An almost single-minded concentration on people – their jobs, their equipment, their personal development – accounts for much of its success.

☐ The quality control process in a decentralized oil-field services business involves careful selection, development, assignment, and compensation of employees working under varying conditions and in widespread locations where close supervision is impossible. In this prosperous company, the process builds shared values and bonds people together.

☐ An international airline, by paying more attention to market economies than to production scale economies, reduces the average size of its aircraft and increases its net income.

☐ Products introduced since 1982 by a well-known financial service generated 10% of its revenues in 1985. The raw material for these products is data already existing in other forms in the company's vast data base.

These examples give a glimpse of forward-looking management practice. When examined closely, they offer insights into the ideas on which successful competitive strategies have been fashioned in the much-maligned and little-understood service sector.

It's no coincidence that dominant industries have cutting-edge management practices. Some U.S. railroads in the nineteenth century pioneered in divisionalized management of their far-flung systems and in good procurement procedures to support their sizable construction and operational needs. At the turn of the century, basic industries led the way in experimenting with scientific management. Then the rise of the large consumer goods manufacturer, epitomized by the auto industry, spawned concepts of decentralization and a full product line aimed at carefully segmented markets.

> *Whatever your business, services have something to teach.*

Today service industries have assumed the mantle of economic leadership. These industries, encompassing trade, communications, transportation, food and lodging, financial and medical services, education, government, and technical services to industry, account for about 70% of the national income and three-fourths of the nonfarm jobs in the United States. In generating 44 million new jobs in the past 30 years, they have absorbed most of the influx of women and minorities into the work force, softened the effects of every post-World War II recession, and fueled every recent economic recovery.

In view of this leadership role, now is a good time to look at the exemplars in the service sector for insights into ways of boosting productivity and altering competitive strategies. Despite their diversity, leading companies in many service industries display some common themes and practices. And they yield lessons for managers in any sector of business. Let's look first at the way the best service companies are structured.

James Heskett is the 1907 Foundation Professor of Business Logistics at the Harvard Business School. Currently he heads an MBA course, Management Policy and Practice. This article is an outgrowth of his latest book, Managing in the Service Economy *(Harvard Business School Press, 1986).*

Integrated functions

Most goods-producing businesses follow the traditional organizational pattern of separate and equally important marketing and manufacturing functions, with coordinating authority at high levels. Some service businesses do the same thing, but the pattern is much less common in service companies where contact with customers is close, as in retailing, passenger transport, and food and lodging. In these businesses, service is marketed and produced at the same place and time, and often by the same person. Naturally, close coordination between marketing and operations management in these cases, regardless of reporting relationships, is essential.

Integration of marketing and operations is often found at very low levels in these organizations. In fact, more than 90% of all field managers in four multisite service companies surveyed in one study claimed responsibility for operations, personnel, and marketing, could not say which was most important, and paid great attention to each.[1]

Even where operations are buffered from marketing activities in organizations offering little customer-contact service, there are ways to break down the traditional functional barriers. Several years ago, the Chase Manhattan Bank launched an effort to upgrade its nonloan products, improve its external communications and customer service, and make its back-office (production) operations more market based. A weak spot was Chase's international business. In the highly visible "product" of international money transfer, differences of viewpoint between marketing–embodied in the account relations manager in the field–and the back office in New York had frustrated communication. Errors were frequent, a large backlog of inquiries about balances and transactions had piled up, and morale in the operations group was poor.

A study ordered by the executive put in charge showed that headquarters accounted for operational errors in only about one-third of all the inquiries and that the marketing people had little idea what operations could offer the bank's customers. The executive traced the backlogged errors to their sources, often a correspondent bank, and resolved them. He launched a campaign to improve operations staff morale around the theme "We make it happen" and formed a new group, the customer mobile unit, consisting of the bank's most experienced international operations people. The unit visited Chase customers at their businesses to help resolve problems and smooth operations. The executive brought the marketing and back-office people together to talk about ways to improve the flow of information. Perhaps most important, the bank revised reporting relationships so that operations units serving specific market segments reported to both the customer relationship manager and the head of operations–a move that improved functional coordination.[2]

The product manager's job was created in many manufacturing organizations to address the problem of coordinating manufacturing and marketing. But in most cases, product managers have had profit responsibility without the authority to coordinate. Assignment to these positions has been regarded as temporary, which encourages decisions with a short-term orientation.

Because of their importance, the high-contact service company makes a point of developing numbers of marketing-operations managers, often carrying the title of store or branch manager. At hand, therefore, is a large cadre of talent from which the company can draw senior managers already trained for administrative responsibilities.

Strategic service vision

The need of most service organizations to plan as well as direct marketing and operations as one function has led to the formation in leading companies of what I call a strategic service vision. Its elements consist of identification of a target market segment, development of a service concept to address targeted customers' needs, codification of an operating strategy to support the service concept, and design of a service delivery system to support the operating strategy. These basic elements are shaded darker in *Exhibit I*.

A company naturally tries to position itself in relation to both the target market and the competition. The links between the service concept and the operating strategies are those policies and procedures by which the company seeks to maximize the difference between the value of the service to customers (the service concept) and the cost of providing it. This difference, of course, is a primary determinant of profit. And the link between the operating strategy and the service delivery system is the integration achieved in the design of both. These integrative links are shaded lighter in *Exhibit I*.

To see how the strategic service vision works, examine the Hartford Steam Boiler Inspection & Insurance Company. For many years, HSB has been in the business of insuring industrial and institutional equipment. Its market targets are organizations using boilers and related pieces of equipment with high operating risk. It offers the same risk reduction as many other insurance companies but positions itself against the competition by emphasizing cost reduction as well.

Exhibit I **Externally oriented strategic service vision**

Target market segments	Positioning	Service concept	Value–cost leveraging	Operating strategy	Strategy-systems integration	Service delivery system
What are common characteristics of important market segments? What dimensions can be used to segment the market? Demographic? Psychographic? How important are various segments? What needs does each have? How well are these needs being served? In what manner? By whom?	How does the service concept propose to meet customer needs? How do competitors meet these needs? How is the proposed service differentiated from competition? How important are these differences? What is good service? Does the proposed service concept provide it? What efforts are required to bring customer expectations and service capabilities into alignment?	What are important elements of the service to be provided, stated in terms of results produced for customers? How are these elements supposed to be perceived by the target market segment? By the market in general? By employees as a whole? How do customers perceive the service concept? What efforts does this suggest in terms of the manner in which the service is: Designed? Delivered? Marketed?	To what extent are differences between perceived value and cost of service maximized by: Standardization of certain elements? Customization of certain elements? Emphasizing easily leveraged services? Management of supply and demand? Control of quality through– Rewards? Appeal to pride? Visibility and supervision? Peer group control? Involving the customer? Effective use of data? To what extent does this effort create barriers to entry by potential competition?	What are important elements of the strategy? Operations? Financing? Marketing? Organization? Human resources? Control? On which will the most effort be concentrated? Where will investments be made? How will quality and cost be controlled? Measured? Incentives? Rewards? What results will be expected versus competition in terms of: Quality of service? Cost profile? Productivity? Morale and loyalty of servers?	To what extent are the strategy and delivery system internally consistent? Can needs of the strategy be met by the delivery system? If not, what changes must be made in: The operating strategy? The service delivery system? To what extent does the coordination of operating strategy and service delivery system ensure: High quality? High productivity? Low cost? High morale and loyalty of servers? To what extent does this integration provide barriers to entry to competition?	What are important features of the service delivery system, including: The role of people? Technology? Equipment? Layout? Procedures? What capacity does it provide? Normally? At peak levels? To what extent does it: Help ensure quality standards? Differentiate the service from competition? Provide barriers to entry by competitors?

Basic element

Integrative element

HSB concentrates on a few types of equipment and has built a large data base on their operating and performance characteristics. (Manufacturers of the equipment often turn to HSB to get wear and maintenance data.) The information furnishes the actuarial base on which HSB prices its insurance. The company's engineers, who inspect customers' equipment before and after it is insured, are also qualified to give advice on preventing problems and improving utilization rates, and through many years of association they often get very close to their customers. As a service manager of one HSB client told me, "If I tried to replace that insurance contract, my operating people in the plant would let me know about it."

This practice enhances the perceived value of the service to the customer at little extra cost to HSB. Of course, by reducing the risk to the customer HSB can improve its own loss ratio.

HSB has a larger cadre of engineers than any of its competitors. These engineers, in tandem with the big data base, make up a service delivery system that capitalizes on the knowledge of marketing and operating managers at all levels of the organization.

The net result is a strategic service vision (though HSB doesn't use the term) that is highly valued by its customers and very profitable for its provider. It addresses implementation issues as part of the strategic plan, and it requires agreement and coordination among marketing and operating managers throughout the organization.

Inner-directed vision

High-performance service companies have gained their status in large measure by turning the strategic service vision inward: by targeting impor-

Exhibit II **Internally oriented strategic service vision**

Target employee group

What are common characteristics of important employee groups?

What dimensions can be used to describe these employee groups?

Demographic?

Psychographic?

How important are each of these groups to the delivery of the service?

What needs does each group have?

How well are these needs being served?

In what manner?

By whom?

Positioning

How does the service concept propose to meet employee needs?

How do competitors meet such needs?

How are relationships with employees differentiated from those between competitors and their employees?

How important are these differences?

What is "good service" to employees?

Does the proposed service concept provide it?

What efforts are required to bring employee expectations and service capabilities into alignment?

Service concept

What are important elements of the service to be provided, stated in terms of results produced for employees and the company?

How are these elements supposed to be perceived by the targeted employee group?

How are these elements perceived?

What further efforts does this suggest in terms of the manner in which the service is:

Designed?

Delivered?

Value-cost leveraging

To what extent are differences between returns to employees and the level of effort they put forth maximized by:

The design of the service concept?

The design of the elements of the operating strategy?

Job design?

The leveraging of scarce skills with a support system?

The management of supply and demand?

Control of quality through–

Rewards?
Appeal to pride?
Visibility?
Supervision?
Peer group control?

Involving the customer in the delivery of the service?

Effective use of data?

Operating strategy

How important is direct human contact in the provision of the service?

To what extent have employees been involved in the design of the service concept and operating strategy?

How desirable is it to:

Increase employee satisfaction?

Increase employee productivity?

What incentives are provided for:

Quality?

Productivity?

Cost?

How does the strategy address employee needs for:

Selection?

Assignment?

Development?

Evaluation?

Compensation?

Association?

Strategy-systems integration

To what extent are the strategy and the delivery system for serving important employee groups internally consistent?

To what extent does the integration of operating strategy and service delivery system ensure:

High quality?

High productivity?

Low cost?

High morale and "bonding" of the target employee group?

Service delivery system

What are important features of the service delivery system, including:

The role of people?

Technology?

Equipment?

Layout?

Procedures?

What does it require of target employee groups?

Normally?

At peak periods of activity?

To what extent does it help employees:

Meet quality standards?

Differentiate their service from competitors?

Achieve expectations about the quality of their work life?

Basic element **Integrative element**

tant groups of employees as well as customers. In the head offices of these organizations, questions such as those listed in *Exhibit II* are heard often. The questions parallel those in *Exhibit I;* but in asking them about employees, management shows it's aware that the health of the enterprise depends on the degree to which core groups of employees subscribe to and share a common set of values and are served by the company's activities.

The basic elements, shaded darker as in *Exhibit I*, start with the service concept designed with employees' needs in mind. The operating strategy is set to meet these needs in a superior fashion at the lowest cost, a result often achieved through the design of the service delivery system. The integrative elements, shaded lighter, include positioning of a service concept, which it is hoped will lead to low turnover, low training costs, and the opportunity to develop shared goals and values. High-performance service organizations invariably have operating strategies designed to maximize differences between operating costs and value perceived by employees in their relations with the company. And delivery systems designed with the operating strategy in mind can form the foundation for remarkable gains in productivity.

A case in point is the ServiceMaster Company, based in Downers Grove, Illinois, which manages support services for hospitals, schools, and industrial companies. It supervises the employees of customers' organizations engaged in housekeeping, food service, and equipment maintenance. These are services that are peripheral to the customers' businesses and therefore get little management attention.

Many of the people whom ServiceMaster oversees are functionally illiterate. To them, as well as its own managers, ServiceMaster directs a service concept centered around the philosophy stated by its CEO, "Before asking someone to do something you have to help them be something." ServiceMaster provides educational and motivational programs to help these employees "be something."

To its own supervisors the company offers training leading to an ambitious "master's" program taught in part by the chief executive. New responsibilities and opportunities present themselves via the rapid growth of the company, approximating 20% per year, nearly all of it from expansion of existing operations rather than acquisition. Elaborate training aids and a laboratory for developing new equipment and materials enhance the employee-managers' "be something" feeling.

For customers' employees ServiceMaster tries to build the "be something" attitude and improve their productivity by redesigning their jobs and by developing equipment and pictorial, color-coded instructional material. In most cases it is the first time that anyone has paid attention to the service of which these employees are a part. ServiceMaster also holds weekly sessions to exchange ideas and offers educational programs to, among other things, develop literacy. ServiceMaster also recruits up to 20% of its own managers from the ranks in jobs it handles. The service concept clearly is improved self-respect, self-development, personal satisfaction, and upward mobility.

Another company slogan, repeated often, is "to help people grow." When a hospital served by the company decided to hire a deaf person, ServiceMaster's local head didn't object. Instead he authorized three of his supervisors to take a course in sign language.

It should be no surprise that the turnover rate among ServiceMaster's 7,000 employees is low. Further, the turnover rate in organizations it services is much lower than the averages for their industries. And when ServiceMaster takes a job, the productivity achieved by supervised support workers invariably rises dramatically.

Now a billion-dollar company, ServiceMaster had a return on equity from 1973 through 1985 that was the highest of all the largest service or industrial companies in the United States, averaging more than 30% after taxes. It oversees the support service employees for 15 hospitals in Japan, which probably makes it the largest exporter of managerial talent to Japan. According to one ServiceMaster executive, "The Japanese immediately recognize and identify with what we do and how we do it." This company turns its strategic service vision inward with dramatic results.

The vision applied

In addition to building a strategic service vision, the best service companies apply it to customers and to those who deliver the service and oversee its delivery – in new or different ways. From my study of organizations like Hartford Steam Boiler and ServiceMaster, I've gathered a series of lessons useful for service providers to consider. These lessons can furnish goods producers food for thought too.

Rethink quality control

Executives whose careers have spanned service as well as manufacturing agree that reaching a consistently high quality level is tougher in services. In high-contact (so-called high-encounter) services, the interaction between two people or more varies with each transaction. In low-contact services, people many miles from the customer have to rely on their own judgment in handling orders and other transactions and in fielding complaints.

The best companies integrate operations and marketing.

Those who have tried to solve the quality control problem by adding more supervision have found that it limits effectiveness. A service transaction cannot be halted, examined, and recycled like a product.

The most effective approaches to the problem have included restructuring of incentives to emphasize quality, designing jobs to give service providers higher visibility in dealing with customers, and building a peer group to foster teamwork and instill a sense of pride.

One incentive that is often effective in organizations ranging from rapid transit companies to hotels is the employee-of-the-month award – especially if based on customer feedback. Both monetary and nonmonetary incentives have been used successfully. What's more, the cost is low.

Making the person who delivers the service more visible is another technique. In England, at the Lex Service Group's luxury auto dealerships, the customer is encouraged to go directly to the mechanic working on the car. The Shouldice Hospital near Toronto, Canada specializes in the repair of hernias using only local anesthetic – a practice that allows the doctor to talk with the patient during the operation. Defective work is referred to the doctor responsible. The remission rate for hernias performed at Shouldice is less than one-tenth that of the average North American hospital. At Benihana, the U.S. chain of Japanese-style steak houses, the chef cooks at a grill in front of the

Exhibit III **How success builds high-contact services**

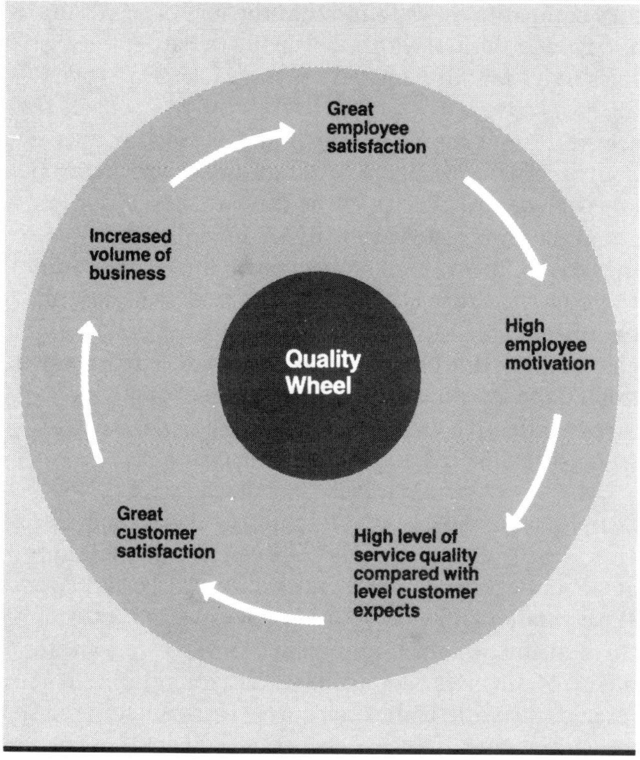

restaurant guests. The chef's visibility and proximity to customers promote a consistently high quality of service and a consistently high level of tips.

Incentives and visibility may be insufficient for those tasks performed without supervision and out of view of the customer. In these cases, some companies rely on careful selection and thorough training of employees and the development of programs to build both a sense of pride in the service and a sense of identification with the company. This bonding process can be hard for rivals to emulate and can thereby contribute to competitive advantage.

Schlumberger's wire-line service has roughly 2,000 geological engineers, each responsible for a mobile rig equipped with more than $1 million worth of computers and electronic gear that helps predict the outcome of petroleum producers' drilling efforts. Each year the company recruits those it considers the brightest of the crop of college engineering graduates, spends months teaching them how to use the equipment, and goes to great lengths to make them feel a part of a special tradition. As one engineer put it recently, "Indoctrination is just as important as technical training." This is all in preparation for an assignment to represent Schlumberger in the field, without direct supervision, often in a remote part of the world. Two measures of the success of this program are Schlumberger's dominant share of the world's wire-line business and the profit-to-sales ratios for this company, which consistently exceed others in its industry in good times and bad.

Often effective in achieving and maintaining quality is peer group control, supported by incentives, training, job design, and service delivery system design. In cases where professional standards have been established for a task, they reinforce peer group control.

In an architectural firm, the mere existence of a policy requiring partners' review of every piece of work can keep partners and associates on their toes. Surgeons are sometimes assigned in teams to foster the learning process and encourage peer group control. A partner of a leading real estate development company told me, "There are three things I'm most concerned about in my work. In this order, they are not to embarrass my colleagues, not to cast a bad light on the company by inadequately serving my clients, and making money." It's not surprising that this company has a strong sense of shared values, reinforced by a policy of encouraging partners to invest in the projects that they propose and carry out.

Recent research suggests that the internal strategic service vision, quality control, and success are connected, especially in those providers of high-encounter service requiring judgment in delivery. I show it as the "quality wheel" in *Exhibit III*. Studies directly link customer satisfaction and the resulting sales volume to the satisfaction derived by the person serving the customer.[3] Naturally, the more motivated the employee, the better the service.

The selection and development of employees, care in assignment, and the layout and equipment of the facility (in a high-contact environment) are all integral elements of the design of the service encounter, which in turn is based on the company's assessment of customer needs. Preconditioning of the customer may also be a part of the design of the service encounter. Review and redesign of the encounter go on continually as the organization assesses how well it is meeting those needs.

A part of the internal service vision is the design of policies and performance measures that further the fulfillment of customers' needs. For example, the server's well-being in the job apparently depends, at least in part, on the extent to which his or her superiors emphasize the solution of problems for customers rather than strict adherence to a set of policies and procedures.[4]

Driving the self-reinforcing elements of the wheel of quality takes a great deal of executive time and requires an honest interest in people across the organization. The senior vice president for finance of Delta Airlines, an organization well regarded for its service and its employee programs, remarked recently, "I would guess that 25% of the time of the finance department officers is spent listening to people problems."

For most service companies, people obviously are more important than machines in the control of quality. But even where the machines employed carry an unusually high value, as in Schlumberger and Delta, developing and building the dedication of people takes precedence.

Reassess the effects of scale

In service organizations, scale economies are often much more important at the company level than at the operating unit level. This is particularly true for companies that have many units over wide areas connected by a common identity. Such scale gives McDonald's and Hertz great purchasing clout and General Cinema the advantage of selling soft drinks of its own manufacture.

Large scale at the company level is important for exploiting network effects, a phenomenon much more important in the service than in the manufacturing sector. To a point, the addition of new network links augments volume for those parts already in place, thus building average network capacity utilization. The addition of service to Las Vegas from its Memphis hub gave Federal Express more volume on the Memphis-New York link. When Visa adds a large retailer to its network of card-accepting establishments, it increases the attractiveness of its credit card to present and potential cardholders and the potential volume to be realized by retailers already accepting the card.

Effective service requires people who understand the idea.

Bigger is not better in those service industries in which the factory must be taken into the marketplace to sell a more accessible, visible, and convenient product that meets customers' needs. Factories operated by the Hyatt and Marriott organizations (called hotels) have not, on average, grown in size for years. These companies have settled on a range of hotel dimensions that can be designed, located, and operated effectively to achieve the capacity utilization, quality of service, and financial performance they want. The range describes sizes at which diseconomies resulting from poor supervision and inflexibility tend to outweigh advantages of larger scale. In the design and siting of hotels, Hyatt and Marriott give the less quantifiable advantages of market flexibility weight equal to operating economies of scale.

At the unit operating level, many service companies have found that the loss of flexibility and greater difficulty in supervising those delivering the service far outweigh any savings realized in operating costs as unit size grows. In the rush to cut costs per seat-mile, for example, many of the world's airlines bought large, wide-bodied aircraft like the Airbus 300 and McDonnell DC-10. While these planes performed admirably, their effective utilization required funneling large numbers of passengers into the airline's hub. Moreover, because business travelers, who represent the most attractive market segment, are prone to choose an airline on the basis of times and frequency of flights, the load and schedule consolidation necessary for effective employment of wide-bodied aircraft worked against the goal of building traffic.

When Jon Carlzon became CEO of Scandinavian Airlines System in 1980, wide-bodied aircraft were used extensively between the airline's hub at Copenhagen and major cities like London and Paris. With smaller DC-9s, SAS funneled travelers between the hub and other Scandinavian cities. To reclaim the business travelers SAS had lost, Carlzon relegated most of the wide-bodies to charter work and offered nonstop flights using DC-9s between Scandinavian and principal European cities.

A size question confronts nearly every power utility in the United States today. For years it was industry gospel that the more power-generating capacity concentrated in one place, the greater the economies of scale. This was the case until the 1970s, when ever-larger units began encountering reliability problems. Furthermore, construction schedule stretch-outs, at times fomented by environmental groups' agitation against big plants, caused the expected power-generating economies to vanish. Finally, an improved capability for transmitting excess energy from one market to another made it possible to buy energy for less than the big units could afford to charge. So many utilities today are meeting the needs of smaller markets' fluctuating demands more economically through new means.

Replace & create assets with information

For decades, manufacturers have sought ways of substituting information for assets. Foremost among these are forecasting and inventory control techniques. For many service operations, information offers creative new ways to substitute for assets.

Heating oil dealers, by maintaining data on the capacity of their customers' tanks, on habitual consumption rates, and on weather, program fuel oil deliveries to provide 100% availability while reducing delivery times and the number of trucks and drivers. These companies substitute information for assets.

The Rural/Metro Fire Department extends effective fire protection at a fraction of the cost of most municipally run fire departments. This Scottsdale, Arizona-based company analyzes data on past fires and uses much smaller, less expensive trucks staffed with smaller crews and equipped with a large-diameter hose that can shoot a lot of water on a fire very fast. On the way to a fire, a truck crew can learn the floor plan of the building to which it is going. While speeding through the streets, the crew examines a microfiche of the layout on a screen. Rural/Metro substitutes information for assets.

Many service industries are information driven, beginning with familiarity between the server and the served. In many (not all), assets have never been allowed to become dominant, perhaps because of limited capital. But with the development of new technologies for processing and communicating information, companies in these industries have advanced far beyond the use of information as a substitute for assets. They are instead using the information they have collected in one business as the basis for new services.

Companies servicing manufactured goods, for example, have built data bases on the types, wear rates, and failure rates of various parts of a furnace, appliance, or automobile. A company can use this information for sending timely service reminders to customers and also to manage parts inventories to reflect the age and condition of the particular machine serviced. In the process, the data have taken on great value for the producers of the goods—and they're willing to pay for the information.

A credit card service builds expenditure profiles for its customers; broken patterns may signal a problem like stolen cards. Theft is sometimes suspected when a large expenditure is made far from the cardholder's address. Instead of outright disallowance of a retailer's request for a big charge, one major travel card issuer tries to determine whether the cardholder indeed is traveling in the retailer's area. Information collected for this service yields person-specific data about travel patterns that often is valuable to airlines and hotel chains (to name two businesses). But the company limits the use of such information to ways that benefit its cardholders.

Dun & Bradstreet's $2.7 billion enterprise is centered on its data base, a file of credit information describing businesses in 30 countries. Through development and acquisition, the file steadily grows. D&B has consistently realized about 10% of its revenues from business that did not exist three years before. Nearly all of these services use the same data base but package the information in different ways. A potential competitor would have to spend an estimated $1 billion—nearly half D&B's net asset value—to duplicate the data base.

Though a data base may constitute a service provider's most important asset, it doesn't appear on the balance sheet and can't be depreciated. But the degree to which many such companies rely on an accumulation of knowledge as their chief competitive weapon and source of new business development suggests opportunities for their counterparts in the manufacturing sector.

Harlan Cleveland has pointed out that information, unlike most manufactured products, is often infinitely expandable (as it is used), compressible, substitutable (for capital, labor, or physical materials), transportable, diffusive (hard to keep secret), and sharable (as opposed to exchangeable).[5] If it is infinitely expandable, those who possess it are limited only by their imagination in creating new ideas, revenue sources, and job opportunities. As the demand for creative exploitation of information grows, so will job creation in the service sector.

The service economy

Many successful service providers have strategies in common that offer lessons to other companies. Among these are:

Close coordination of the marketing-operations relationship.

A strategy built around elements of a strategic service vision.

An ability to redirect the strategic service inward to focus on vital employee groups.

A stress on the control of quality based on a set of shared values, peer group status, generous incentives, and, where possible, a close relationship with the customer.

A cool appraisal of the effects of scale on both efficiency and effectiveness.

The substitution of information for other assets.

The exploitation of information to generate new business.

Why these particular lessons among all I might cite? For one reason, they feature characteristics that distinguish many service industries from goods-producing industries. Notice the emphasis on people, ideas, and information instead of things. For another, they promise twin benefits as part of a business strategy. Each can further differentiation of the service product as well as lower costs.

These lessons have significance for the economy too. While the service economy has wrought a gigantic social restructuring of the United States, it has come in for unwarranted criticism for its low rate of productivity gains. Companies like those I have described, however, have created new jobs while raising productivity. If other companies learn these lessons, job opportunities in the service sector will continue to expand and productivity continue to rise. These developments will ease the pressures for the inflation of service prices, sharpen the already respected competitiveness abroad of U.S.-based services, and contribute to the partnership between services and manufacturing that is crucial to a healthy, balanced national business base.

References

1 Christopher H. Lovelock, Eric Langeard, John E. G. Bateson, and Pierre Eiglier, "Some Organizational Problems Facing Marketing in the Service Sector," in *Marketing of Services*, ed. James H. Donnelly and William R. George (Chicago, Ill.: American Marketing Association, 1981), p. 168.

2 See James F. Loud, "Organizing for Customer Service," *The Bankers Magazine*, November-December 1980, p. 41.

3 Benjamin Schneider and David E. Bowen, "New Services Design, Development, and Implementation and the Employee," in *New Services*, ed. William R. George and Claudia Marshall (Chicago, Ill.: American Marketing Association, 1985), p. 82; and Eugene M. Johnson and Daniel T. Seymour, "The Impact of Cross Selling on the Service Encounter in Retail Banking," in *The Service Encounter*, ed. John A. Czepiel, Michael R. Soloman, and Carol F. Surprenant (Lexington, Mass: D.C. Heath, 1985), p. 243.

4 This is the implication of John J. Parkington and Benjamin Schneider in "Some Correlates of Experienced Job Stress: A Boundary Role Study," *Academy of Management Journal*, June 1979, p. 270.

5 Harlan Cleveland, "Information as a Resource," *The Futurist*, December 1982, p. 37.

Reprint 87206

Will services follow manufacturing into decline?

James Brian Quinn and Christopher E. Gagnon

"Daily we encounter the same inattention to quality, emphasis on scale economies, and short-term orientation that earlier injured manufacturing."

Although they probably know better, many executives still think of the service sector in terms of people making hamburgers or shining shoes. These images belie the complexity, power, technical sophistication, and economic value of activities that now account for more than 68% of the nation's GNP and 71% of its employment. Worse, they help perpetuate a set of myths about service industries that lead managers and policymakers to ignore their full potential. Worse still, such inattention and complacency threaten to undermine the competitive ability of these industries at a time when their importance to the national economy has never been greater.

Services are actually all those economic activities in which the primary output is neither a product nor a construction. Value is added to this output by means that cannot be inventoried–means like convenience, security, comfort, and flexibility–and the output is consumed when produced. *Exhibit I* suggests the scale and power of services in the United States.

Now more than ever, we need a vital service sector. To suffer the beating there that we have already endured in manufacturing would be a national disaster. Even so, we dither and deny while warning signs appear. It is no small matter that net positive trade balances in services have fallen steadily since the beginning of the decade–from \$41 billion in 1981 to \$21.4 billion in 1985. Of the total, business services declined more than \$9.3 billion between 1981 and 1985, when the net balance was a meager \$0.2 billion. The most serious losses have occurred in travel and transportation-related industries. Although U.S. airlines have maintained a steady 38% to 40% of the world's revenue passenger miles during the past decade, its once powerful international carriers, Pan Am and TWA, have fared poorly at the hands of foreign competitors like Swissair and JAL, which have made heavy long-term investments in their fleets and paid close attention to the quality of care given passengers.

Mr. Quinn is the William and Josephine Buchanan Professor of Management at Amos Tuck School, Dartmouth College. He is an author and a consultant on technological and strategic management. Mr. Gagnon of the Amos Tuck School was principal research associate for the project on which this article is based.

The threat to other services is real and immediate. These confrontations are not limited to markets abroad; we can lose over here too. Many U.S. markets for services are no safer from foreign competition than were domestic markets for manufactured goods. Indeed, foreign direct investment in the U.S. service sector has exploded since the mid-1970s, although the rate of increase has slowed a bit in recent years. It is sobering to realize that many of the great names in services–names like Twentieth Century-Fox, Stouffer's Hotels and Restaurants, Intercontinental Hotels, Saks Fifth Avenue, Gimbels, Marshall Field, Spiegel, A&P, Grand Union, and Giant Food–have foreign owners now. Even Goldman-Sachs has a new partial owner from abroad.

U.S.-based services offer even more opportunity for growth–and for mismanagement–than product-based industries did in the past. While there is still time, it is essential to take a hard look at how we think about services, how we manage them, and how much they contribute to the nation's economic health.

Facing the facts

Many views of services do not jibe with the facts:

☐ Contrary to popular opinion, services are not responses to marginal demands that people sat-

Exhibit I **Components of U.S. gross national product**

	Billions of 1985 dollars
Total GNP	**$ 3,998**
Agriculture, forestry, and fishing	$ 92
Mining and construction	305
Manufacturing	796
Total products sector	**$ 1,193**
Transportation and utilities	$ 268
Communications	109
Wholesale trade	277
Retail trade	375
Finance, insurance, and real estate	627
Other services	639
Total private services sector	**$ 2,295**

Source:
U.S. Department of Commerce, Bureau of Economic Analysis. Revised figures, August 1986. The $510 billion difference between the sum of these two subtotals and GNP is predominantly government expenditures.

The study

This article is an early report on a multiphase study. To date, we have collected and analyzed data from the Bureau of Labor Statistics, the U.S. Department of Commerce, Compustat, and PIMS. We have also interviewed experts in the financial services, communications, health care and delivery, retailing, airline, and rental car industries. Some interesting initial patterns have emerged from the data we have assembled and the anecdotes we have collected.

isfy only after they meet their product needs. People value services at least as highly as they do manufactured products and purchase them in much less cyclical patterns than manufacturers.

☐ Service companies produce value added at comparable or higher rates than do product companies.

☐ The service sector is at least as capital-intensive as the manufacturing sector, and many industries within it are highly technological.

☐ Service industries tend to be as concentrated as manufacturing and to have companies of sufficient scale to be sophisticated buyers and even producers of technology.

☐ Service industries lend themselves to productivity increases great enough to fuel continuing real growth in per capita income.

Services substitute directly for manufactured products across a wide spectrum. Few customers care whether a computer accomplishes a function by a hardware circuit or by its software. If anything, they may prefer the software approach–especially when, as with CAD/CAM, it substitutes for production machinery at a fraction of the cost. In much the same way, self-service at retail levels can lower real costs for manufactured goods as much as can savings in the production process itself.

Such substitutions boost productivity and value added in real, concrete terms. In fact, our data indicate that measured value added in the service sector is at least as high as in manufacturing. (See *Exhibit II.*) Moreover, services are less cyclical than manufacturing. In the last two decades, employment in services has advanced on average 2.1% during economic contractions and 4.8% during expansions. In contrast, in the goods-producing sector, employment has declined an average of 8.3% during recessions and increased only 3.8% during expansions. (See *Exhibit III.*) These figures mean that people will stop buying many products before they will sacrifice service essentials like education, telephones, banking, health care, and police and fire protection.

Nor, as the evidence shows, is the service sector less labor-intensive or technologically based than manufacturing. Capital stock per service worker has been rising since the mid-1960s and has already surpassed that for manufacturing workers, and some service industries–notably rail and pipeline transportation, communications, public utilities, and air transport–are among the most capital-intensive of all industries.[1] In perhaps the most complete study of relative capital intensities in U.S. industry, nearly half the top 30 industries–out of 145 surveyed–were services.[2] (*Exhibit IV*, based on the profit impact of management strategy [PIMS] data base, confirms these findings.)

The service sector is also a major market for high technology. The most recent studies available show that some 80% of the computing, communications, and related information technologies equipment sold in the United States during 1982 went to the service sector and that in Britain 70% of all computer systems sold in 1984 went to the service sector.[3] Contrary to the current stereotype that the service sector is much too small and diffuse to buy large technological systems or to do research on its own, PIMS data suggest that concentration in the service sector is comparable to that in manufacturing and that, at the margin, services are about as attractive to capital as manufacturing is. (See *Exhibit V* and *Exhibit VI.*)

Service industries have both the financial power and the structural rationale to buy technology as needed. Large banks, insurance companies, airlines, utilities, communications companies, hospitals, and retail chains all have the scale not only to purchase technology but also to help manage its conception, design, and development.

In turn, these combined infusions of capital and technology can raise levels of service productivity and per capita national income ever higher. According to the latest available studies, between 1975 and 1982 there was a 97% increase per service worker in new technology investment, which resulted in many service industries achieving significant productivity improvements.[4] In others, where improvements have not yet been impressive, the potential for productivity-enhancing automation remains quite high. (See *Exhibit VII.*)

Exhibit II **PIMS value added**

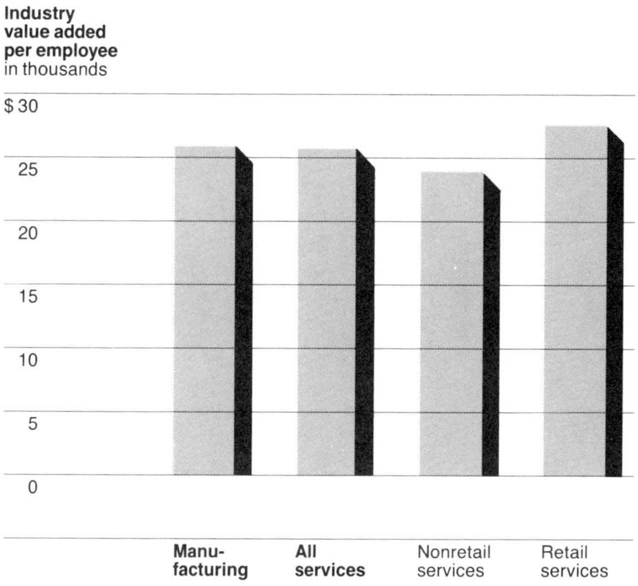

Note:
This index is derived from an industry survey and calculations from the Profit Impact of Management Strategy (PIMS) data base, Strategic Planning Institute, Cambridge, Mass.

New terms of competition

Although the effects of deregulation complicate the picture, most experts agree that new technologies have restructured the service industries extensively in recent years. Unfortunately, these technologies have also made the industries vulnerable to the same modes of attack that earlier so rapidly undermined the nation's manufacturing capability. If executives understand the relevant technical and competitive forces, they can take advantage of some powerful opportunities to boost service performance and help develop switching and entry barriers that shut out foreign competitors. If executives do not comprehend these potentials or if they act with the complacency of manufacturing managers a decade ago, they will leave major portions of the U.S. economy wide open to foreign incursions.

What restructurings has technology brought? And what are their implications as opportunities or points of attack?

Economies of scale

Every service industry we studied reported important new economies of scale driven by the application of new technologies. The first-order effect in most cases was a revised competitive structure characterized both by increased concentration and by increased fragmentation, niching, or segmentation. To realize the potential scale economies of new technologies, many intermediate and large companies in each service industry merged to form giant enterprises. But within each industry smaller companies also identified local niches – or specialized service needs – and concentrated successfully on these.

In the rental car industry, for example, the large companies, in conjunction with the airlines, could initially lock up virtually any long-distance traveler with their instantaneous guaranteed reservation systems. They even had international subsidiaries or affiliates connected electronically. Soon, niche operators proliferated under this pricing umbrella. Automated computer and telephone answering devices allowed both "superelegant" and "Rent-A-Wreck" extremes to serve local markets with office-in-the-home operations, and midsize operations like Alamo and Agency could segment target groups with particular needs like cut-rate and conference rentals.

Automation of the securities trading process, to cite another example, changed the entire structure of that industry in the mid-1960s and early 1970s. Under the old paper-based system, shares traded

Exhibit III **Recession resistance of service sector**

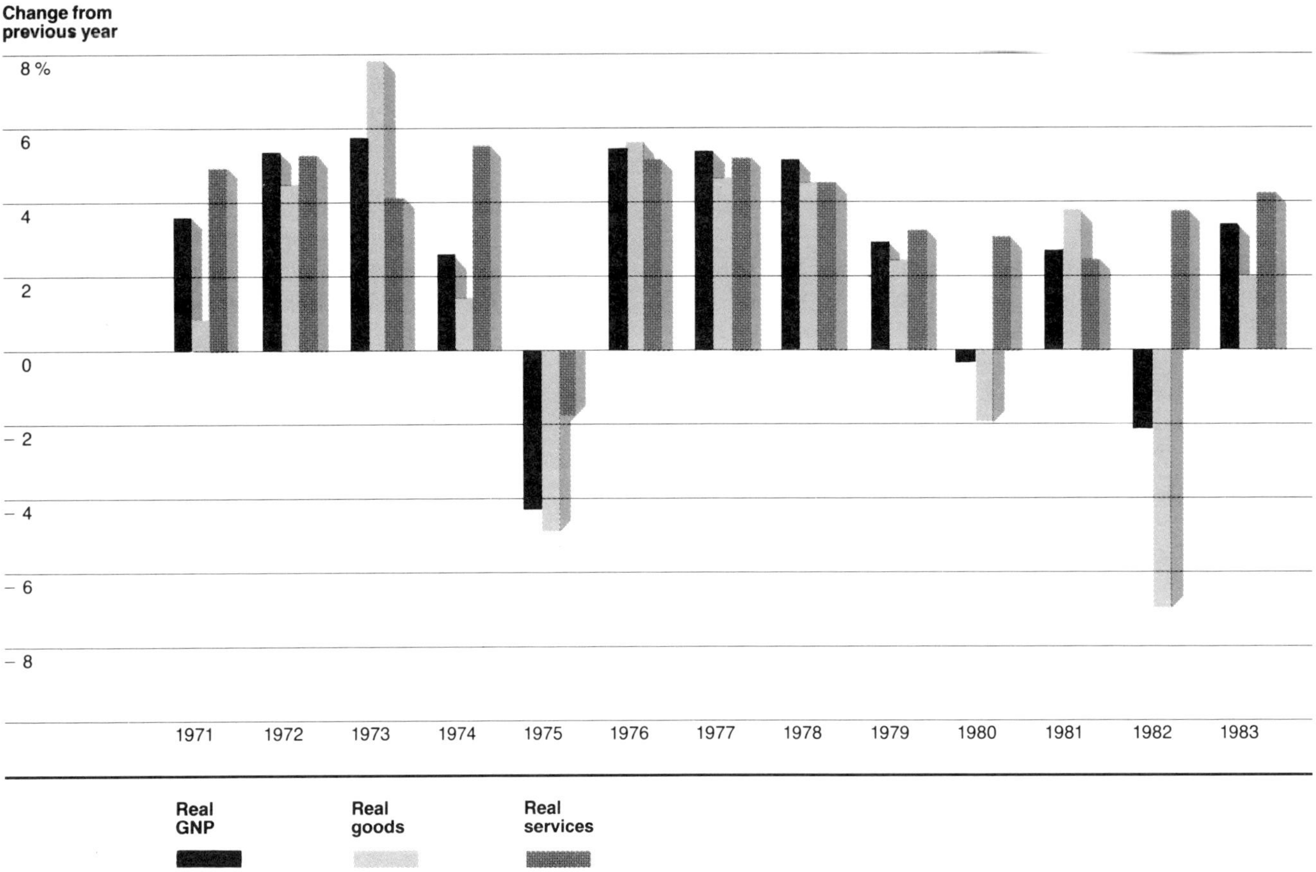

Source:
U.S. Department of Commerce, *Statistical Abstract of the United States*, 1984.

had to be physically delivered from the seller's agent to the buyer's. As daily volumes approached 12 million shares, only the big banks could hire and manage enough people to keep track of their securities trades each day. Smaller firms began to fail because they could neither control nor process their securities in a timely fashion.

Finally, Wall Street firms formed Central Certificate Services (later the Deposit Trust Company), which brought essentially all securities certificates under one roof. Rather than having to move shares physically, a single set of accounting entries could change their ownership. After five or six years, the system became totally electronic and smaller brokers could tie into the depository. Today, such automated clearinghouses handle virtually all private and government transactions.

New technology has also had scale effects on medical care. Because of the reimbursement system first in use, the original effect of capital-intensive technologies was to centralize treatment in the larger hospitals. Small practitioners and hospitals did not have enough patients to be able to afford constant updating of their diagnostic, surgical, and recovery equipment. This centralization led specialized medical practices to form around the biggest hospitals and spurred the creation of regional referral centers to handle particularly difficult cases. As a result, many smaller hospitals suffered, closed down, or joined cooperative networks with the larger centers.

The average cost of an inpatient stay soared from $729 in 1972 to $2,898 in 1984, when diagnosis-related group restrictions began to shift cost patterns. With patient care growing ever more impersonal in the large centers, new and less costly distribution systems have begun to emerge along both vertical (home care, primary care, specialty care) and horizontal (pediatrics, obstetrics, dermatology, internal medicine) axes. Such complex alternative systems as electronically linked HMO networks have grown almost ten times in number since 1971, and the number of people they serve from 3.1 million to more than 13.6 million. Ambulatory surgical centers have exploded from 400 in 1982 to 1,200 today.

Management of these complex service systems has become so critical a factor in the health care industry that large private companies have found it profitable to apply their skills to both hospitals and other parts of the system. To obtain scale economies, insurance companies have integrated forward into health care management and some hospitals have begun to offer insurance.

Exhibit IV **Capital intensity**

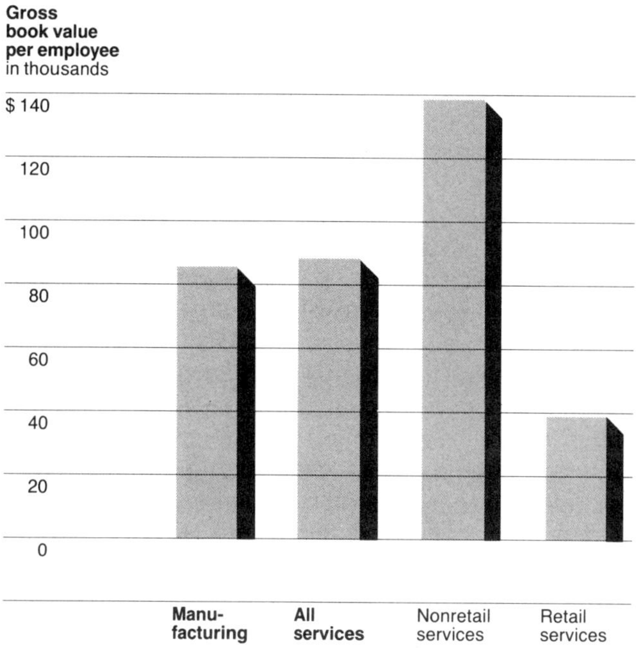

Source:
Calculated from PIMS data base, Strategic Planning Institute, Cambridge, Mass.

Economies of scope

Beyond its important effects on scale, technology often creates powerful second-order effects – economies of scope – that allow entirely new service products to move through established networks or systems with little added cost. Once debugged, communications and information-handling technologies permit the distribution of a much wider set of services to a more diffuse customer base, yet they often actually lower marginal costs on old products as equipment, development, and software investments are allocated over a broader base of applications. In addition, such new technologies frequently offer further strategic benefits in terms of more rapid new-product introduction or faster response to competitors' moves.

When Federal Express started its overnight small-package delivery business, it developed advanced computer communication systems to link its couriers' trucks with dispatchers. To prevent inflight delays, it installed the most up-to-date civilian flight-control system available. To track, control, and handle its packages and to manage its operations efficiently, Federal created its COSMOS electronic network. With these major subsystems in place, the company could, as opportunities arose, add overnight document delivery, large packages, international services, and finally, two-hour ZapMail document delivery, with costs lower than those of competitors.

Insurance companies first automated their backroom activities in the mid-1960s when their industry was stable and heavily regulated. Handling premium billings and collections brought them dramatic productivity gains. When wildly fluctuating interest rates hit the industry a decade later, the companies had to alter their products rapidly to attract new premiums and to offset the effects of customers borrowing against their policies at low interest rates. In such an environment, only companies with flexibly designed computer and control systems could design or deploy their products quickly enough to get a competitive edge. Previously, insurance companies had brought out new rate books only every three to five years. Successful industry executives said that without effective electronic and software systems, they could neither have conceived of the variety of new products needed in the 1970s nor have explained and introduced them to their widespread agent and customer bases.

Many smaller insurance companies, which could not afford the huge costs of sophisticated electronic networks, attest to this fact. They failed, sold out, merged, or concentrated on localized or specialized services falling outside the interests of larger companies.

Given both scale and scope economies as well as the relatively low expense of transporting services internationally, large technologically sophisticated U.S. companies like Citicorp and a deregulated AT&T should have powerful international advantages. But there is no reason to be complacent. The same is true for certain foreign competitors. Large Japanese and European banks have already expanded into U.S. markets. NEC has carved out a potent position at the crucial interface between computers and communications. And large European and Asian distribution companies have proved they have the talent and resources

Exhibit V **Concentration in service sector**

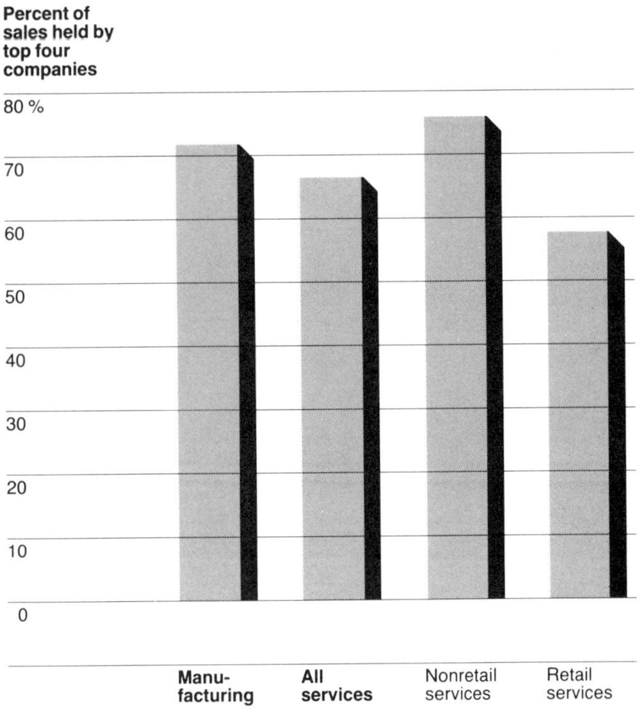

Source:
PIMS data base, Strategic Planning Institute, Cambridge, Mass.

to acquire and manage underperforming U.S. retail networks.

Complexity

Technology in services permits the management of new levels of complexity. Law firms, for instance, can in a matter of hours complete more exhaustive background searches, prepare more intricate contracts, and document resulting settlements more thoroughly than they previously could in weeks. Computer models and data base networks allow specialists of all sorts to analyze much more complex problems. So powerful are some of these technologies that in some research fields, they can identify critical relationships and pose new hypotheses, not merely assist in analyzing and testing data. For instance, computerized models can now suggest feasible structures for complex but as yet unknown proteins, and gene machines can manufacture them within hours for initial testing against potential antigens or biological insults.

On a broader scale, the velocity of transactions in monetary exchanges and commercial activity has increased so much that without well-developed electronic support systems, it is impossible to compete effectively in many markets. For example, annual world trade in products and services is perhaps $2.5 trillion to $3 trillion, but Eurodollar transactions alone aggregate some $75 trillion. A major stock exchange is open somewhere in the world all but six hours in the workweek, and these markets are instantly accessible from almost any point on the globe.

The shortened transaction times, heightened volatility, and sheer complexity inherent in these transactions simply overwhelm older modes of doing business. But the technologies that make such systems possible also have their own side effects, which further undermine established terms of competition, strategic management, and even economic policy-making. International capital markets are so closely tied together that it may be impossible for one country to create cost-of-capital advantages for its companies through its policy actions. Control of national economics may become equally problematic.

Within companies, the effects of complexity reinforce economies of both scope and scale. Companies can deliver better and more varied services with no significant cost penalties. They can simultaneously achieve a high degree of segmentation in their activities and lower their costs. Computerized reservation systems allow airlines to analyze their costs and their customers' buying behavior in such detail that they can optimize margins on each type of demand and meet each competitor's response. The resulting crazy quilt of prices leads customers to concentrate more on services provided, which offers the major carriers further opportunities to segment with services like specialized meals, wheelchairs, committed seat assignments, luggage verification, and even counseling for nervous passengers. Yet within a wide range of activities, marginal costs can actually drop as systems overheads are allocated over a wider base.

Similarly, bar-code scanners give retailers instant feedback on their sales and inventory movements–hence opportunities for much more control over inventory, margins, and even operating costs. Some retailers now go a step farther and sell this information to help suppliers and manufacturers fine-tune their own strategies. Their sophisticated information systems enable major chains to stock "superstores" with a variety of products that were previously uneconomical to keep under one roof and to branch out into retail drug, home center, mail order, or small specialty chains that may have higher value-added potentials.

Exhibit VI **Investment efficiency**

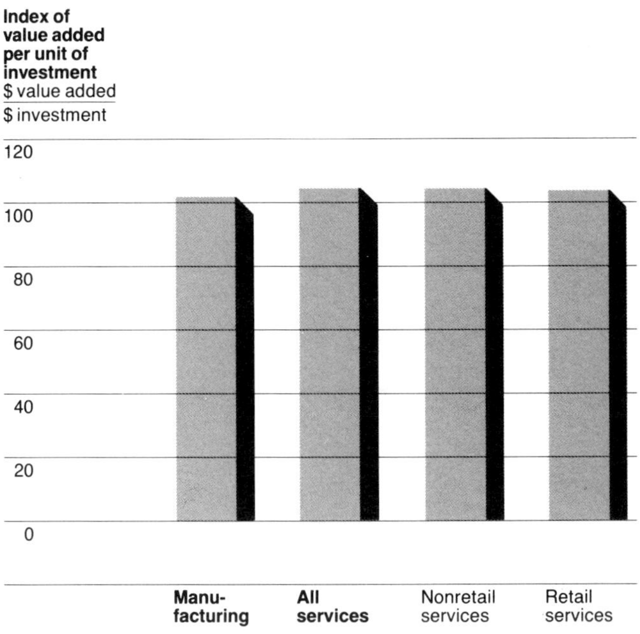

Source:
Calculated from PIMS data base, Strategic Planning Institute, Cambridge, Mass.

Exhibit VII **Productivity increases in service sector**

	Percentage of average annual improvement	
	1960-1983	**1970-1983**
Telephone and communications	6.1 %	6.8 %
Air transportation	5.8	4.5
Railroad (revenue traffic)	5.1	4.8
Gas and electric utilities	2.7	1.0*
Commercial banking		0.9†
Hotels and motels	1.6	0.8

*1981 data.

†1982 data.

Source:
Bureau of Labor Statistics, Office of Productivity and Technology.

Boundary crossing

Still another effect of technology on competition in services is to break down many of the traditional boundaries among these industries. Some of the most obvious examples are in financial services. Many consumers now use their banks and brokers almost interchangeably because neither is seriously restricted in its scope of operations and consumers no longer care which intermediaries handle their investments. As a consequence, banks, insurance companies, and brokerage houses increasingly offer a similar range of financial products and services and compete head to head on the basis of location, knowledge, and cost.

At the same time, retailers and manufacturers like GM and Ford have also parlayed off their strengths in consumer credit to present similar financial service offerings. Their actions have eroded the distinction between financial services and product marketing to the point that the two are inseparable. In fact, General Motors Acceptance Corporation's $75 billion in assets and $1.02 billion in profits (more than one-fourth of the parent's earnings in 1985) have made it the nation's largest single holder of consumer debt.

The implications of information services technologies for competitive strategy in product-oriented companies like Exxon and Boeing are equally profound. These companies' profits can be made or devastated by how well they develop and deploy knowledge about supplier costs, new technologies, exchange rates, changing regulations, swap potentials, and political and market sensitivities. In the aggregate, more money is often made in the goods sector through information and services than through production activities. In fact, one study shows that about three-quarters of the total value added in the goods sector is created by *service* activities within the sector.[5]

Services and production are becoming so widely substitutable that distinctions between the sectors seem more arbitrary than helpful. Executives and policy analysts often view activities like product design, market research, accounting, and data analysis as product costs if they are carried out within manufacturing concerns but as services if they are provided externally. Internal salespeople are classified under manufacturing employment, but external sales representatives and wholesalers are called service providers. If a farmer harvests his own grain, the costs are production costs; if the farmer hires a professional combine operator, the activity is a service.

Measuring the service component of international trade is even more complicated. If a U.S. bank has a loan officer in its British subsidiary who relies heavily on analyses developed in New York for a deal consummated in London, is this an export? A return on investment abroad? Or just a local sale of services? In manufacturing, how much of a company's overseas income is due to its products rather than the

Exhibit VIII **Productivity of service sector in United States and Japan**

		Value of output per labor hour* Japan 1970	Japan 1980	United States 1970	United States 1980	United States as a percentage of Japan 1980
Private domestic business		$ 3.59	$ 6.01	$ 9.40	$ 10.06	167 %
Agriculture		1.37	2.38	16.53	18.36	771
Selected services	Transportation and communications	3.86	5.66	9.29	13.14	232
	Electricity, gas, and water	14.01	19.74	21.98	25.38	129
	Trade	2.88	4.53	6.88	7.92	175
	Finance and insurance	6.69	12.03	8.21	8.20	68
	Business services	3.39	3.60	7.69	7.59	211
Manufacturing		3.91	8.00	7.92	10.17	127

*In constant 1975 dollars.

Source:
Measuring Productivity: Trends and Comparisons, from the First International Productivity Symposium, Tokyo, Japan, 1983 (New York: UNIPUB, 1984).

embodied technology, training, systems, or service support the company extends from its U.S. base?

International competitiveness

Perhaps the most crucial impact of service technologies will be on international trade. True, in 1980 worldwide service exports were only about 3.7% of GDP and those of the United States were only 1.4%. But worldwide service trade was growing at a 19% rate–much greater than world production's 14% growth rate.[6] In 1985, some 18% of the world's trade was in measurable services, whereas 49% was in manufactures and 33% in other goods.

First, service and support technologies related to products contribute much of their total value. Second, services have also vastly restructured the international manufacturing marketplace. Communication technologies, of course, permit manufacturers to coordinate their design, sourcing, distribution, and manufacturing strategies on a worldwide basis to minimize costs. But other services have affected manufacturers' costs and global trade patterns even more.

Cheaper and more flexible transportation systems have loosened geographic constraints on production and encouraged the transfer of skills, technology, and manufacturing knowledge among countries. Containerization and new onboard storage techniques (like that for liquefied natural gas) have made extensive world trade possible for such things as fragile high-technology goods, volatile chemicals and gases, and certain bulk items like coal. Computerized navigation and satellite communications systems have lowered transportation risks and insurance costs and allowed instant redeployment of in-transit inventories like oil to maximize margin gains.

Some consequences, though equally far-reaching, are less clear. In earlier years, the improvement of service-related technologies permitted manufacturing to migrate from industrialized countries to developing areas, where labor or materials costs were lower. In the near future, however, these same technologies may place an even higher competitive premium on staying physically close to the marketplace to respond more rapidly to customers' service needs. This situation could lead to the "remanufacturing" of Western countries.

For example, at GM, Saturn's planned five- to eight-day cycle from order to delivery of an options-loaded vehicle may make it difficult for foreign plants to compete. The same may be true of just-in-time inventoried products throughout the U.S. production system. Japanese auto companies building cars in the United States are already bringing their suppliers here for just this reason. In the years ahead, manufacturing may actually return to the older industrialized economies because of the services needed to satisfy customers in a highly diffuse marketplace, where rapid and flexible responses create high-value premiums.

The inescapable fact is that services have become a critical cost dimension of a nation's manufacturing competitiveness. Services can dramatically raise or lower the real cost of producing goods domestically – through differences in transportation, communications, financing, insurance, and health care costs. Efficiency in these sectors also improves the real wealth of laborers for any given wage level and thus relieves upward cost pressures. While the Japanese have been superb in certain mass manufacturing fields, their productivity (in terms of GNP per person) has consistently lagged behind that of the United States, largely because we have a more productive service sector. (See *Exhibit VIII.*) Properly used, U.S. services offer our manufacturers unique international competitive advantages.

Sowing for the harvest

The United States has painfully learned that its goods-producing industries are vulnerable to challenge from abroad. So too are the services on which much of the nation's economic health depends. Technology-driven changes in how these industries compete provide great opportunities, to be sure, but they also open the door to new threats. It is difficult for well-run service establishments to maintain their competitive advantages when everyone can buy the same hardware and software and connect into the same networks and when suppliers of these new technologies have strong incentives to sell their products as widely and quickly as possible. Seemingly secure markets may be easy to invade.

The resulting losses could be tragic. On the one hand, our policies of deregulation in communications and financial markets may have given us a lead that, some say, other countries' more regulated sectors will find it difficult ever to close. But there is little room for complacency. Deregulation also opens domestic U.S. markets to foreign competition. And many countries and foreign companies have proved that their skills in managing service enterprises are formidable indeed – note only the Swiss PTT and Japan's NTT, Oberoi Hotels, Swissair and Singapore airlines, Barclays Banks and Lloyd's Insurance, and the spreading influence of the Japanese tourist and banking industries.

It will take hard and dedicated work not to dissipate our broad-based lead in services, as we did in manufacturing. Many of the same causes of lost position are beginning to appear. Daily we encounter the same inattention to quality, emphasis on scale economies rather than customers' concerns, and short-term financial orientation that earlier injured manufacturing. Too many service companies have been slow to invest in the new market opportunities and flexible technologies available to them. They have stayed with their old concepts too long and have concentrated on cost-cutting efficiencies they can quantify rather than on adding to their product value by listening carefully and flexibly providing the services their customers genuinely want. Haven't we heard this once before?

The cost of losing this battle is unacceptably high. If service industries are properly nurtured, they will grow and generate much of America's future wealth. If they are misunderstood, disdained, or mismanaged, the same forces that led to the decline of U.S. manufacturing stand ready to cut them to pieces. With some 70% of the U.S. economy already in services – not including the three-fourths of all manufacturing costs that represent support services – the stakes are immense. It can happen here.

References

1 Stephen Roach, "Information Economy Comes of Age," *Information Management Review*, January 1985, p. 9.

2 Ronald Kutcher and Jerome Mark, "The Services Sector: Some Common Misperceptions Reviewed," *Monthly Labor Review*, April 1983, p. 21.

3 Richard Kirkland, "Are Service Jobs Good Jobs?" *Fortune*, June 10, 1985, p. 38; and "Information Makes the Money Go Round," City of London survey, *The Economist*, July 6, 1985, p. 5.

4 *U.S. National Study on Trade in Services* (Washington, D.C.: Office of the U.S. Trade Representative, December 1983), p. 24.

5 Ibid., p. 21.

6 Ibid., pp. 13, 16; "The World is Their Oyster: A Survey of International Banking," *The Economist*, March 16, 1985, p. 5; *Euromoney Yearbook 1983* (London: EuroMoney Publications Ltd., 1983), sec. 1.3.

Author's note: We greatfully acknowledge the support of the Bell & Howell and Bankers' Trust companies.

Reprint 86611

The hospital as we know it is too costly, too unwieldy, and too inflexible to survive.

A Radical Prescription for Hospitals

by Jeff Goldsmith

Declining margins, excess capacity, mature product portfolio, bureaucratic overburden, poorly planned and executed diversification moves, rapid CEO turnover – a familiar litany of symptoms of U.S. industries in trouble. But this list describes not the airlines, retailers, basic manufacturing, real estate, or financial services; it describes the $200 billion American hospital industry.

The seriousness of these symptoms has only recently become obvious. In 1985, despite an almost 20% decline in inpatient use nationwide, hospitals enjoyed record operating margins. By 1987, they had almost doubled their 1980 revenues. By the end of last year, however, operating earnings had evaporated. Administrator turnover had reached an annual rate of 25% to 30%, and trustees (along with new management teams) have been left to struggle to keep their institutions on an even keel.

> The new hospital will reach out into homes and residential communities.

Hospitals face a grim short-term prognosis. After moderating in the 1984 to 1986 period, health costs are soaring again. Health insurance premium increases of 20% to 40% have prompted employers to renew their scrutiny of benefit costs. Congress and the Bush administration, looking for ways to pare the huge budget deficit, seem likely to take the knife to Medicare, hospitals' largest payment source. The unrelenting economic pressure on hospitals may finally produce the long-predicted closure of the 700 to 1,000 institutions in deepest trouble. But the shakeout of excess capacity will mean better margins for the survivors.

The Medicare cutback would come at a time when the elderly are pressing for expanded benefits. Future expansion, however, may in large part be paid for through a reduction in provider payments. For the federal government, meanwhile, control over physician and ambulatory service costs looms as the likely prime cost-containment issue. Just as the government imposed prospective payment limits on inpatient care in the early 1980s, it has begun to put similar restrictions on outpatient care, notably surgery

Jeff Goldsmith is national adviser on health care to Ernst & Whinney, the accounting firm, and chairman of Integrated Medical Campus, an ambulatory health facilities development company. He also lectures at the Graduate School of Business, University of Chicago, and operates Health Futures, Inc., a Chicago consulting firm.

and diagnostic services. These limits may reward freestanding providers with lower overhead and punish hospitals that have remained less efficient than the competition.

While the short-term outlook is not promising, the long-term view is in a sense very favorable. Except for major regional institutions, the acute-care hospital as we know it will probably not survive. The successful hospital of the future will be a far different institution that focuses on early diagnosis and management of chronic illness. The hospital of the future will reach out into homes and residential communities as much as it depends on the sick to cross its threshold for help. This is the radical prescription in store for hospitals.

More Than a Business

Alert administrators and trustees have learned that hospitals have to be run more like businesses. Yet hospitals are vastly more complex than a typical business. Techniques and strategies that may work in manufacturing and retailing have proved inadequate in the hospital sector.

The common business technique of achieving economies through larger scale, for example, has not yet proven effective in hospitals.[1] Scale economies in health care delivery have been confined mainly to supply and capital costs, but they represent only about 15% of total costs. Personnel represents the biggest cost item, 60% or more of the total. Large multihospital systems, nursing home chains, and health maintenance organizations so far have been unable to extract scale economies in the use of their key resource, skilled professionals and technicians. (A notable exception is the Kaiser Foundation Health Plans, which have not only economized in the use of doctors but also have effectively leveraged "physician extenders" like nurse practitioners.)

Those who have predicted that American health care would be dominated by a handful of large, vertically integrated organizations have come to realize that there is little that is meaningful about health care which is national. Health services are a neighborhood business, beginning and ending with a doctor and a patient.

Far from gaining economies, many large health care organizations became bloated by building up corporate staffs and layers of management. These added little value to patient service. Indeed, the greater organizational complexity and slower decision making alienated workers at the bottom of the pyramid who were dealing with patients – and possibly contributed to the exodus of nurses, technicians, and other patient care workers from the labor market. Today health care institutions are plagued by surpluses of white-collar workers and shortages of skilled care givers.

The investor-owned chains, the organizations that were supposed to dominate the hospital scene by exploiting economies of scale and access to equity capital, have had mixed results. In 1987, after several successful years, seven chains either declared bankruptcy or suspended debt-service payments. The four largest (Hospital Corporation of America, Humana, National Medical Enterprises, and American Medical International) divested marginal facilities and pruned shaky diversification ventures.[2]

The chains that remained profitable did so by adopting aggressive rate increases and by taking positions in psychiatric and rehabilitation services – which will be targets of future cost scrutiny. While the larger companies' profitability has rebounded somewhat in the wake of the restructuring, smaller chains have limited prospects. Many are selling back part or all of their holdings to their physicians through limited partnerships.

Diversification is another business move that has not been successful for hospitals. They have maintained control of two vital growth markets, ambulatory surgery and high-tech home health care (like parenteral nutrition and infusion therapy). Often the hospitals entered these fields via joint ventures with doctors or suppliers. But excursions into more remote fields frequently ended in financial disaster. The major investor-owned hospital management companies, for example, plunged into sponsorship of captive health insurance plans. Today, after those companies have written off hundreds of millions of dollars in losses, only Humana remains committed to the business.

In the nonprofit hospital sector, zany acquisitions in unrelated businesses sapped resources and diverted management's attention. In one major case, poorly run diversification led to insolvency and subsequent merger into a large organization. This was Scioto Memorial Hospital in Portsmouth, Ohio, whose diversification activities included a bottling company, a Montana dude ranch, and a life-care center in a distant community whose collapse ultimately brought down the facility.

Another business strategy is to bypass the "middleman" and deal directly with the consumer. Hos-

1. Stephen Shortell documents this fact in "The Evolution of Hospital Systems: Unfulfilled Promises and Self-Fulfilling Prophecies," *Medical Care Review*, Fall 1988, p. 177.

2. See Regina E. Herzlinger, "The Failed Revolution in Health Care – The Role of Management," HBR March-April 1989, p. 95.

"Watch out for Pynchotte. He's getting haircuts on a level far above his present standing in the company."

pitals' effort to "retail" health services by circumventing the physician yielded little and posed a competitive threat to them.

Although many patients were attracted to the fixed-price packaging of some services, like emergency care and obstetrics, hospitals that relied on a retail strategy were among the worst casualties of the decade. Instead of trying to correct deficiencies in facilities, equipment, and medical staff, they relied on marketing to fill beds. Some of the smaller management chains, such as Republic Health Corporation, tried to gloss over their deficiencies with slick merchandising strategies like media campaigns and brand name identification.

This experience reinforced a truth: doctors remain at the center of care delivery. Their share of the health care dollar rose in the 1980s in proportion to the hospitals' decline. Organizations like the HMOs that set out to "tame" their physicians suffered bloody defeat in initial skirmishes. It is probably not coincidental that health care organizations dominated by doctors–regional multispecialty group practices, university teaching hospitals, and high-tech regional hospitals–were the big winners in the market share struggle of the 1980s. These institutions relied on long-established referral and consultative networks with medical colleagues to arrange patient flow from as far away as a few hundred miles.

The big winners in the HMO shakeout were those like Kaiser Permanente that are run by powerful physician groups. The large, multispecialty doctor group is an extremely potent competitive force in most major markets and seems likely to retain or increase that power even with the impending physician payment reforms.

Out of these experiences a valuable business lesson emerged: health care organizations cannot be managed by simply manipulating the revenue structure. Many of the strategies the institutions pursued to market their "products" did not work. These included advertising campaigns, new-business development aimed at corporate or retail markets, cutting prices to stimulate volume, marketing excess capacity through captive insurance plans, and deals with doctors (acquisition of practices, paid directorships–often thinly disguised bribes–to woo them from competing hospitals, and favors like below-market rents and health club memberships). The strategies produced neither measurable market share nor incremental earnings gains.

During the 1980s, hospitals suffered from a classic case of "marketing myopia," throwing dollars after a maturing acute-care market while ignoring a crucial shift in the demand for care. Demand for inpatient services plummeted in this decade because technological changes and new practice patterns made much custodial care unnecessary.

Most hospitals, however, continued to measure their performance by their inpatient census and directed their marketing efforts to filling empty beds. Selling empty beds at a discount to health insurance plans, forgiving the Medicare deductible for stays by elderly inpatients, and advertising amenities of inpatient care (even lobster dinners and candlelight suppers!) were attempts not to meet new needs but to leverage existing capacity.

By concentrating on the revenue side of operations and on traditional–and costly–institutional care, administrators and their boards ignored the possibility of fundamental changes in hospitals' product and philosophy.

A New Mission

Hospitals are creatures of the Industrial Revolution. They were intended to be warehouses for unfortunates dying of tuberculosis, smallpox, pneumonia, and other infectious diseases. Owing to the advent of public health measures and the invention of antibiotics, by the late 1940s most infectious diseases had come under control. Over the next 30 years, hospitals adapted to a different mission: treating the symptoms of heart disease, cancer, and other chronic illnesses. Today, however, advances in the diagnosis and treatment of chronic diseases are undercutting that mission.

Our contemporary concept of disease has been shaped by the acute infection – a crisis brought about by an external agent that throws our bodies into violent disequilibrium. Chronic illness, however, is universal, progressive, and inherent in the aging process. As we age, our organs gradually lose their ability to cope with stress and external threats, so we become more susceptible to such diseases as cancer, diabetes, and Alzheimer's. Today, despite the threat of viral illnesses like AIDS and hepatitis, 80% of all deaths and 90% of all illness are related to chronic illness.

> The central business truth is that doctors remain at the center of health care delivery.

A person growing old can have heart disease, diabetes, or Parkinson's for many years before it poses a threat to life. Our contemporary care system, hospitals and health insurance alike, tends to ignore the disease until it reaches that life-threatening stage. Then legions of technicians and screaming ambulances (even helicopters!) rush in to fight it. Addressing the symptoms of chronic illness after long neglect (with transplantation of damaged organs) and coping with trauma and failure have emerged as the hospital's principal missions.

While chronic illnesses are, by definition, incurable, they can be deflected from life-threatening manifestations by timely and accurate diagnosis, advanced drugs and technologies, and lifestyle changes. More elegant diagnostic technologies, from magnetic resonance imaging and ultrasound to genetic probes, are enabling doctors to detect the onset of chronic illness early enough to pursue effective therapy. Managing the course of chronic illness and restoring the already compromised patient to improved functioning is the emerging mission of the health care system.

A shift in the hospital's focus away from tardy treatment of chronic illness and toward diagnosis and management requires bringing services to the patient as much as bringing the patient to the hospital. In 20 years, health care will be concentrated in the home and in the residential community, not on the hospital campus.

Demand for acute care seems likely to shrink as a consequence of these changes. Inpatient services will be increasingly reserved for a smaller number of patients suffering from trauma and multiorgan failure. Inpatient care will be just too expensive for any other use. On the other hand, chronic care will require day treatment facilities, user-friendly ambulatory services, and an ability to reach into the home through care givers and electronic monitoring. Even with an aging population, the need for inpatient care will lessen enormously as doctors detect and control chronic illnesses earlier in their progression.

Technology and shifting practice patterns are transforming the contemporary hospital into the critical-care hub of a dispersed network of medical and social services spread across the community and knit together by computer networks and health insurance contracts. In the years ahead, ambulatory services like diagnostic imaging, laboratory testing, and emergency and unscheduled treatment will constitute the hospital's principal product. Indeed, in many smaller facilities they already do.

Because of technological advances, these services can be delivered anywhere in the institution's service area, not necessarily on campus. Focusing the organization's resources on acute care carries with it the opportunity cost of failing to invest in the services and information systems needed to create this dispersed network.

The Next Five Years

To reach that stage intact, today's hospital must weather the next wave of economic pressures. It must survive in an environment of narrowing cash flows, deteriorating credit, and critical scarcity of technical staff. A successful strategy will require:

Renewed and deepened collaboration with physicians.
Solutions to the productivity problem.
Refocusing ambulatory and chronic-care services.
Managing for medical value.

I'll take up these elements in turn.

Collaboration with Doctors. Hospitals have rediscovered, to their dismay, that doctors are central to their well-being. The government's freeze on Medicare payments to hospitals has squeezed their cash flows. And the abrupt rise in physician use of the hospital for Medicare patients has tightened the squeeze. Every measure of intensity of use by Medicare recipients – length of stay, laboratory and imaging procedures, intensive care, and operating room time – has risen in most hospitals. From 1985 to 1987, radiology and laboratory use for elderly patients rose by more than 8%, but Medicare did not make up the extra costs.[3] This burden threatens spreading red ink for treating the elderly, who make up the hospital's most important constituency.

And when administrators seek doctors' cooperation in controlling this use, they are often told, "But my patients are too sick." Translated, this protest means either "go away" or "prove to me my treatment pattern is inappropriate given my patients' needs." That's a justifiable challenge, and most hospitals regrettably lack information systems that can relate resources employed to the severity of the patient's illness.

After the introduction of diagnosis-related groups (DRGs) in 1983 by a cost-conscious federal government, hospitals were pleasantly surprised by their physicians' willingness to examine and alter treatment patterns to cope with the threat of fixed Medicare payments. Several years of profitability, however, dissipated the sense of urgency that had attended the appearance of the stricter Medicare rules. And hospital managements and physicians drifted apart on other important issues.

But now, against a backdrop of frozen Medicare payments, massive underwriting losses for insurers, and renewed growth of managed-care plans, hospitals and doctors must reinvigorate their collaboration and control how the hospital's resources are to be used in treating patients. Medical staffs must use hospital services more sparingly or administrators will lack the flexibility to make meaningful cost reductions. The most critical (and controllable) services are radiology and laboratory services and intensive care units, where excessive charges may exceed the stringent Medicare limits.

> Hospitals will bring services out to patients as much as patients to the hospital.

Furthermore, the expansion of ambulatory care systems and decentralized service networks cannot proceed unless doctors become involved. In many communities, such as Dallas and Los Angeles, doctors have ignored the hospitals and established their own ambulatory and chronic-care services, including "surgicenters" and freestanding reference laboratories.

But now, with the impending federal restrictions on payments to physican-owned enterprises – and the likely deterioration of doctors' credit in the face of fee reductions and overcrowding in the M.D. field – physicians will need hospitals to create ambulatory services as much as hospitals need them.

Productivity Improvement. Health care managers and physicians can anticipate several years of tightening restrictions on payments as the health economy reacts to surging cost pressures. Deepening discounts to private health plans and the continued failure of Medicare to keep pace with escalating costs have already created cash flow problems for the hospital industry.

A lack of meaningful cost information has hampered the industry's response to this pressure. But at least as important has been a lackadaisical attitude toward labor productivity. Because they are so labor intensive, cost-conscious hospitals (indeed, all health services) have to look first for more effective ways to use employees.

In the early 1980s, many hospitals anticipated economic pressures and modestly reduced their work forces. When prosperity ensued unexpectedly, however, many hospitals loosened the reins on staffs. As a consequence, full-time employment per occupied bed rose more than 20% from 1980 through 1987.[4] About one-third of this increase was in registered nurses. In the name of improved continuity and quality of care, administrators let nurse's aides and licensed practical nurses go and replaced them with more costly baccalaureate nurses.

A high fraction of the remaining growth in employment came in staff functions like marketing, finance, and information systems. These were also highly trained and therefore expensive additions. The ratio of white-collar to line workers – meeting goers to care givers – grew to worrisome proportions.

In the next few years, hospitals will be simplifying their structures and eliminating layers of administrative cellulite built up in the prosperous 1980s. Indeed, the recent earnings rebounds of the investor-owned hospital chains reflect in part their success in cutting overhead and staffing costs. But it took as many as three years of sustained effort. Current practice notwithstanding, not every specialty in the institution will have its own department. More professional and technical employees will be cross-trained as hospitals try to reverse the rigid specialization that has plagued them in recent years.

Even in this computer age, hospitals are often driven by manual information systems clogged with paper. Maintenance of the paper medical records con-

3. Prospective Payment Assessment Commission, "Medicare Prospective Payment and the American Health Care System," Report to the Congress (Washington, D.C., U.S. Government Printing Office, June 1988).

4. American Hospital Association hospital panel survey. Figure is adjusted to reflect stepped-up outpatient activity.

New Diagnostic and Treatment Paths Bypass the Hospital

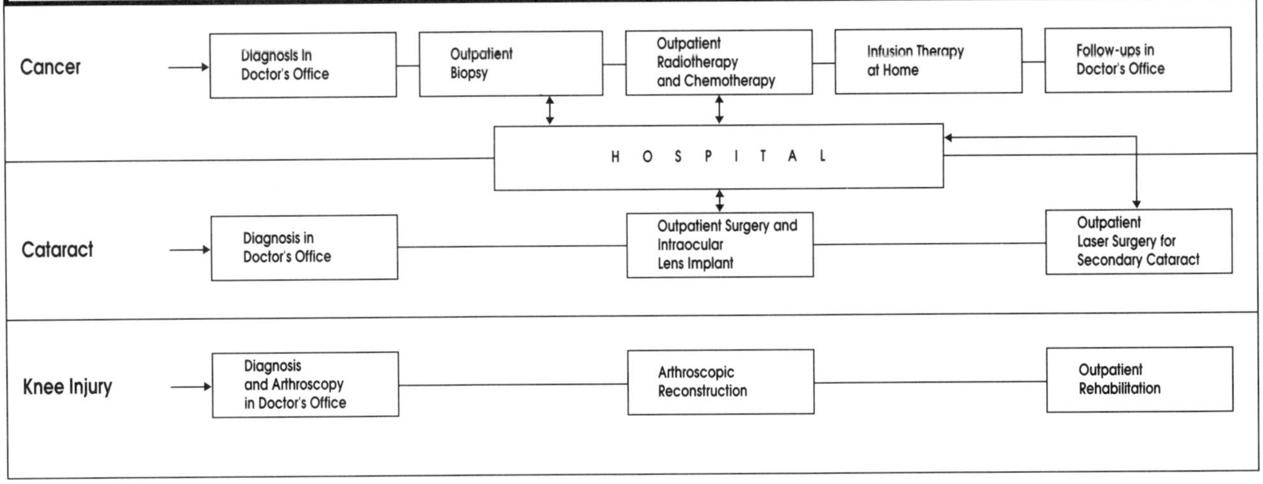

sumes one-third of scarce nursing time on the patient floor. Like many facilities, Harris Methodist Hospitals in Fort Worth are installing bedside terminals to relieve the demands of paperwork on nurses and furnish the necessary clinical information without resorting to a chart.

Hospitals also are moving to real-time electronic transmission and settlement of medical claims, which eliminates the need for an army of clerical workers generating and settling paper bills. Hospital productivity adequate to cope with late twentieth century pressures cannot be gained with operating systems from Dickens's London.

Refocusing of Services. In this decade, hospitals have wisely staked out positions in two big health-care growth markets, ambulatory surgery and home care for the chronically ill. Most administrators, however, have viewed success in these areas as a means to an end – producing more inpatient volume in one case and fewer inpatient-related losses in the other – rather than an end in itself. Emphasis on these services in their own right, and organization of them to weather the impending cost-containment pressures, is an urgent task.

Ambulatory services are the fastest growing part of the hospital; already, ambulatory visits outnumber inpatient admissions in some smaller facilities by as much as 10 to 1. And yet at most hospitals, outpatient diagnostic and surgical services remain stepchildren. In some larger institutions, they are scattered like anemones in a coral reef. In a teaching hospital, a cancer patient on a treatment visit may be obliged to walk more than a mile through a maze of corridors to reach the admission, laboratory, and radiology stations and an examination room. Ambulatory services are located where it is convenient for the inpatient-focused bureaucracy, not for the outpatients or the people dealing with them.

By mid-1989, some 80% of Medicare outpatient services – surgery and diagnostic radiology – will be subject to payment limits similar to those imposed on inpatient care. The ceilings will reward the free-standing and independent providers, because of their lower overhead, at the expense of the hospital. Hospitals that cannot make their outpatient units more cost-effective will find themselves losing money in their fastest growing sector. Constraints set by private insurers are sure to follow.

The expansion of outpatient services is only the most obvious element in the evolution of the hospital plant. Increasingly, treatment paths for complex illness are bypassing the hospital. Whole surgical disciplines have disappeared from inpatient suites, and often from the hospital altogether. These include ophthalmology and plastic surgery as well as major segments of gynecologic surgery, general surgery, and orthopedics. The accompanying chart illustrates three typical situations.

New, noninvasive or minimally invasive diagnostic tools, like Doppler ultrasound and flexible, fiber-optic scopes, are changing the way physicians practice internal medicine. Gastroenterology and cardiology are migrating into outpatient and nonhospital settings. As the diagram shows, cancer treatment, including sophisticated radiotherapy and chemotherapy, is now mostly outpatient and detachable from the hospital campus.

Hospitals are taking a number of steps to cope with the changing situation. They are relocating ambulatory services out of the inpatient areas, in some cases still on campus but in buildings devoted to this use. Often these structures also contain doctors' offices, so they serve the dual purpose of raising the number of M.D.s practicing on campus and upgrading ambulatory diagnostic and surgical care. St. Vincent Medical Center in Portland, Oregon and Baptist Medical Center in Oklahoma City have erected elegant buildings following this plan.

Some hospitals are moving ambulatory services to areas remote from the plant and establishing so-called integrated campuses. These combine outpatient diagnostic and surgical services with doctors' office space, while the hospital continues to provide inpatient and critical care backup. Like the on-campus facilities, some of the integrated campuses are owned jointly with physician groups. One of the most ambitious campuses is Western Reserve Health System's in Youngstown, Ohio, where 22 acres contain doctors' offices, ambulatory diagnostic services, community health education facilities, and a nursing center specializing in Alzheimer's disease. A possible addition is a freestanding birthing center.

The ambulatory facilities are designed to make more effective use of patients' and doctors' time while offering much of the sophisticated technology formerly reserved for inpatients. The facilities are staffed in a lean fashion consistent with the emerging cost constraints.

A refocusing of chronic care also requires hospitals to alter their managerial philosophy. Hospital managements have viewed chronic-care services as a way to dump patients once their acute care is over. Now management must view them as part of a continuum of services. Not surprisingly, hospitals' interest in this type of care got a stimulus from cost limits set by Medicare. Under the DRG restrictions, hospitals often found themselves seeking novel ways to handle patients once the costs of care had reached the Medicare ceilings.

> Home medical treatment is a booming industry that one-third of the nation's hospitals offer.

One way is to cooperate with doctors to develop systems for diagnosing or screening patients for chronic illness. Once the diagnosis is made, the effort turns to treatment, often involving initiative by the patient and the patient's family. Another route is cooperation with community agencies to establish day treatment programs in rehabilitation and nutritional and drug therapy. Memorial Sloan-Kettering Cancer Center in New York City has a day facility for treatment of people who need long-term chemotherapy and other complicated services that do not require overnight stays.

Many services need not be situated on the hospital campus but in neighborhoods where patients can reach them easily. When public transportation is lacking, some hospitals have set up transport networks.

Many of the chronically ill, of course, can receive treatment in their own homes. Most older people with heart disease and diabetes are in this category. Home medical treatment is a booming industry, and about one-third of the country's hospitals now offer it. For example, a hospital-based nurse visits a bedridden elderly woman to administer a premixed bag of medication.

Chronic care forms a bridge between traditional medical services and social services, which health insurers (including the federal government) historically have refused to reimburse. So hospitals that venture into this area are taking on some risk. But as the Medicare program shifts to an emphasis on chronic care, which it is likely to do in the next five years, and as insurance mechanisms like social HMOs emerge to span the gap between acute and chronic care, the risk will diminish.

The promise of more cost-effective health care delivery rests on community-based ambulatory and chronic-care services. Hospitals that approach development of these innovations as ends in themselves, instead of vehicles for boosting occupancy rates, will be ready to meet the competition, adjust to squeezes in care payment systems, and respond to the needs of their communities.

Manage for Value. As cost constraints get tighter and tighter, managing in a world of scarce resources without damaging the quality of patient care becomes a more pressing issue. Discussion of quality is long overdue in an industry that too often assumes that quality is "the way we used to do things" in the days before the industry was forced to become accountable for costs. The real issue buried in the emerging quality debate is not quality per se, but value. The issue is whether a certain medical procedure benefits the patient, given its cost. Or could it be eliminated, and the expenditure saved, without compromising the medical outcome?

The lack of economic accountability in medicine is easy to explain, if not excuse. There has been no objective standard for measuring the value of treatment. Nor has there been a way for those who pay for care to compare the value of services furnished by one provider with those of another.

5. See Robert B. Hughes, Sandra S. Hunt, and Harold S. Luft, "Effects of Surgeon Volume and Hospital Volume on Quality of Care in Hospitals," *Medical Care,* June 1987, p. 489.

By the mid-1990s, employers and the government will have systems that measure the efficacy of various treatment patterns and will also be able to compare the value of providers' services. The avid public consumption of even limited (and often misleading) quality indicators like mortality rates for surgical procedures suggests that a ready market exists for such information.

> By the mid-1990s, it will be possible to compare the value of different hospital services.

Some quality indicators, such as the prevalence of anesthesia deaths or nosocomial (hospital-borne) infections, are manageable and therefore present opportunities for improving patient care. The most important quality measures, however, are connected to clinical outcomes and track the welfare of patients after they leave the hospital. The quality signs include reduced readmission rates, years of symptom-free living, improved productivity on the job, and job attendance. Analysts are also developing methodologies to relate the severity of a patient's illness to the appropriateness of the care he or she is receiving.

Accurate quality measurement is critical to reducing the cost of hospital treatment without compromising the patient's health. It can help hospitals and doctors develop clinical protocols to manage a patient's illness according to its seriousness and eliminate wasteful clinical practices. What may come out of these efforts is a *negative* correlation between quality and cost–that is, elegant and conservative treatment–and sparing use of expensive technologies when they can be avoided, thus producing greater benefit to the patient at less cost.

A growing body of evidence links frequency of surgical performance to better outcomes.[5] A recent study by the inspector general of the Department of Health and Human Services established that concentration of Medicare coronary bypass-graft surgery in the highest volume medical centers would save the program almost $200 million while improving mortality rates. Hospitals have found that doctors who seldom come upon a given medical condition tend to use more resources to treat the patient and often produce a poorer outcome.

Hospitals and their medical staffs have a choice. They can develop their own means of measuring and controlling quality or they can stand by and watch as the government and private payers impose standards on them. They can set and enforce their own standards of appropriate care or they can let the courts define their standards via malpractice litigation. Hospitals that decide to manage the value of a medical encounter will have a huge advantage in an increasingly sophisticated and discerning medical marketplace.

Heading in New Directions

However well hospitals meld their objectives and operations with those of their doctors, however efficiently they use their work forces, and however conscious they become of delivering high-quality service, the fact remains that the facility as it has existed for generations is a dinosaur. It is too costly, too unwieldy in its single setting, and too inflexible.

In the future, acute care will be concentrated in a small number of high-tech regional centers treating traumatic and chronically ill patients. Community hospitals will continue to provide some acute care, like obstetrical services and surgery for victims of illness. Still, most of this care will be ambulatory and often located off campus.

Increasingly, the chronically ill will be treated in the home or in settings remote from the traditional hospital facility. As technological advances continue to strike at disease, most illness will be associated with the infirmities of aging. The community hospital will decentralize its services and weave them into the fabric of the neighborhood and the community. The main role of most community hospitals will be diagnosis and treatment of the chronically ill.

As the American population ages, we will be reallocating resources from acute to chronic care. Those hospitals that recognize this shift in societal priorities can plan to adapt their programs and facilities to meet these new needs.

Reprint 89306

Strategies for Customer Relations

To learn how to keep customers, track the ones you lose.

Zero Defections: Quality Comes to Services

by Frederick F. Reichheld and W. Earl Sasser, Jr.

The *real* quality revolution is just now coming to services. In recent years, despite their good intentions, few service company executives have been able to follow through on their commitment to satisfy customers. But service companies are beginning to understand what their manufacturing counterparts learned in the 1980s–that quality doesn't improve unless you measure it. When manufacturers began to unravel the costs and implications of scrap heaps, rework, and jammed machinery, they realized that "quality" was not just an invigorating slogan but the most profitable way to run a business. They made "zero defects" their guiding light, and the quality movement took off.

Service companies have their own kind of scrap heap: customers who will not come back. That scrap heap too has a cost. As service businesses start to measure it, they will see the urgent need to reduce it. They will strive for "zero defections"–keeping every customer the company can profitably serve–and they will mobilize the organization to achieve it.

Customer defections have a surprisingly powerful impact on the bottom line. They can have more to do with a service company's profits than scale, market share, unit costs, and many other factors usually associated with competitive advantage. As a customer's relationship with the company lengthens, profits rise. And not just a little. Companies can boost profits by almost 100% by retaining just 5% more of their customers.

Defecting customers send a clear signal: profit slump ahead.

While defection rates are an accurate leading indicator of profit swings, they do more than passively indicate where profits are headed. They also direct managers' attention to the specific things that are causing customers to leave. Since companies do not hold customers captive, the only way they can pre-

Frederick F. Reichheld is director of Bain & Company and leader of the firm's customer-retention practice. W. Earl Sasser, Jr. is a professor at the Harvard Business School.

vent defections is to outperform the competition continually. By soliciting feedback from defecting customers, companies can ferret out the weaknesses that really matter and strengthen them before profits start to dwindle. Defection analysis is therefore a guide that helps companies manage continuous improvement.

Charles Cawley, president of MBNA America, a Delaware-based credit card company, knows well how customer defections can focus a company's attention on exactly the things customers value. One morning in 1982, frustrated by letters from unhappy customers, he assembled all 300 MBNA employees and announced his determination that the company satisfy and keep each and every customer. The company started gathering feedback from defecting customers. And it acted on the information, adjusting products and processes regularly.

As quality improved, fewer customers had reason to leave. Eight years later, MBNA's defection rate is one of the lowest in its industry. Some 5% of its customers leave each year–half the average rate for the rest of the industry. That may seem like a small difference, but it translates into huge earnings. Without making any acquisitions, MBNA's industry ranking went from 38 to 4, and profits have increased sixteenfold.

The Cost of Losing a Customer

If companies knew how much it really costs to lose a customer, they would be able to make accurate evaluations of investments designed to retain customers. Unfortunately, today's accounting systems do not capture the value of a loyal customer. Most systems focus on current period costs and revenues and ignore expected cash flows over a customer's lifetime. Served correctly, customers generate increasingly more profits each year they stay with a company. Across a wide range of businesses, the pattern is the same: the longer a company keeps a customer, the more money it stands to make. (See the bar charts depicting "How Much Profit a Customer Generates over Time.") For one auto-service company, the expected profit from a fourth-year customer is more than triple the profit that same customer generates in the first year. When customers defect, they take all that profit-making potential with them.

It may be obvious that acquiring a new customer entails certain one-time costs for advertising, promotions, and the like. In credit cards, for example, companies spend an average of $51 to recruit a customer and set up the new account. But there are many more pieces to the profitability puzzle.

To continue with the credit card example, the newly acquired customers use the card slowly at first and generate a base profit. But if the customers stay a second year, the economics greatly improve. As they become accustomed to using the credit card and are satisfied with the service it provides, customers use it more and balances grow. In the second year–and the years thereafter–they purchase even more, which turns profits up sharply. We found this trend in each of the more than 100 companies in two dozen industries we have analyzed. For one industrial distributor, net sales per account continue to rise into the nineteenth year of the relationship.

As purchases rise, operating costs decline. Checking customers' credit histories and adding them to the corporate database is expensive, but those things need be done only once. Also, as the company gains experience with its customers, it can serve them more efficiently. One small financial consulting business that depends on personal relationships with cli-

How Much Profit a Customer Generates over Time

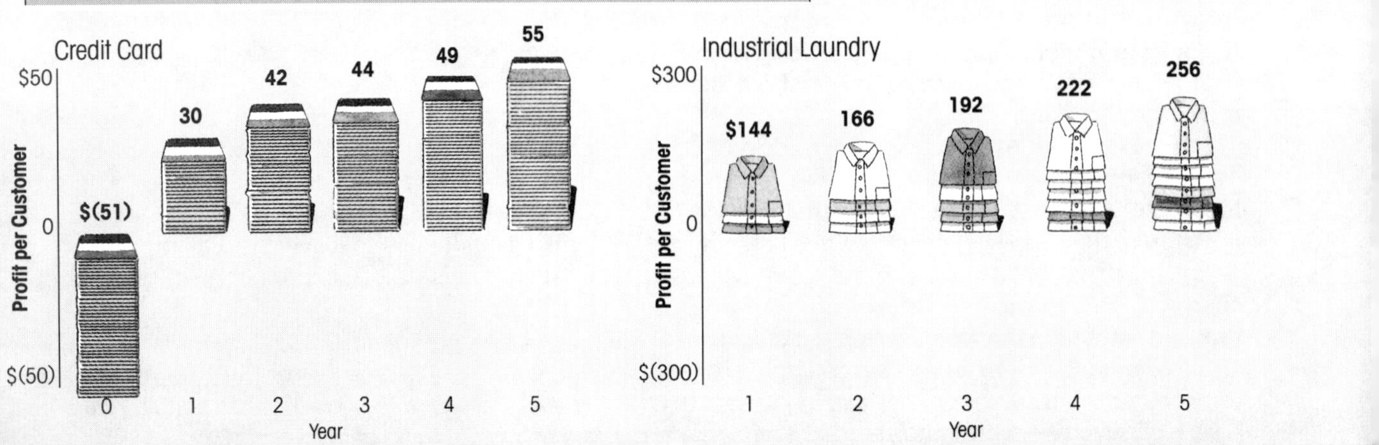

ents has found that costs drop by two-thirds from the first year to the second because customers know what to expect from the consultant and have fewer questions or problems. In addition, the consultants are more efficient because they are familiar with the customer's financial situation and investment preferences.

Also, companies with long-time customers can often charge more for their products or services. Many people will pay more to stay in a hotel they know or to go to a doctor they trust than to take a chance on a less expensive competitor. The company that has developed such a loyal following can charge a premium for the customer's confidence in the business.

Yet another economic boon from long-time customers is the free advertising they provide. Loyal customers do a lot of talking over the years and drum up a lot of business. One of the leading home builders in the United States, for example, has found that more than 60% of its sales are the result of referrals.

These cost savings and additional revenues combine to produce a steadily increasing stream of profits over the course of the customer's relationship with the company. (See the chart "Why Customers Are More Profitable over Time.") While the relative importance of these effects varies from industry to industry, the end result is that longer term customers generate increasing profits.

To calculate a customer's real worth, a company must take all of these projected profit streams into account. If, for instance, the credit card customer leaves after the first year, the company takes a $21 loss. If the company can keep the customer for four more years, his or her value to the company rises sharply. It is equal to the net present value of the profit streams in the first five years, or about $100.

When a company lowers its defection rate, the average customer relationship lasts longer and profits climb steeply. One way to appreciate just how responsive profits are to changes in defection rates is to draw a defection curve. (See the graph, "A Credit Card Company's Defection Curve.") This shows clearly how small movements in a company's defection rate can produce very large swings in profits.

The curve shows, for example, that as the credit card company cuts its defection rate from 20% to 10%, the average life span of its relationship with a customer doubles from five years to ten and the value of that customer more than doubles–jumping from $134 to $300. As the defection rate drops another 5%, the average life span of a customer relationship doubles again and profits rise 75% –from $300 to $525.

> Customers who defect to the competition can tell you exactly what parts of the business you must improve.

The credit card business is not unique. Although the shape of defection curves vary across industries, in general, profits rise as defection rates fall. Reducing defections by just 5% generated 85% more profits in one bank's branch system, 50% more in an insurance brokerage, and 30% more in an auto-service chain. (See the chart "Reducing Defections 5% Boosts Profits 25% to 85%.") MBNA America has found that a 5% improvement in defection rates increases its average customer value by more than 125%.

Understanding the economics of defections is useful to managers in several ways. For one thing, it shows that continuous improvement in service quality is not a cost but an investment in a customer who generates more profit than the margin on a one-time sale. Executives can therefore justify giving priority

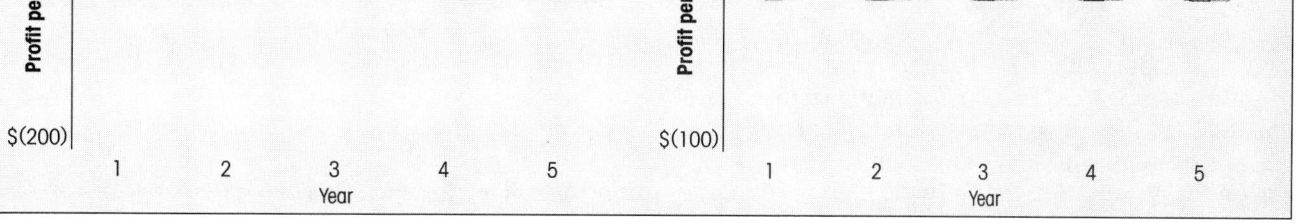

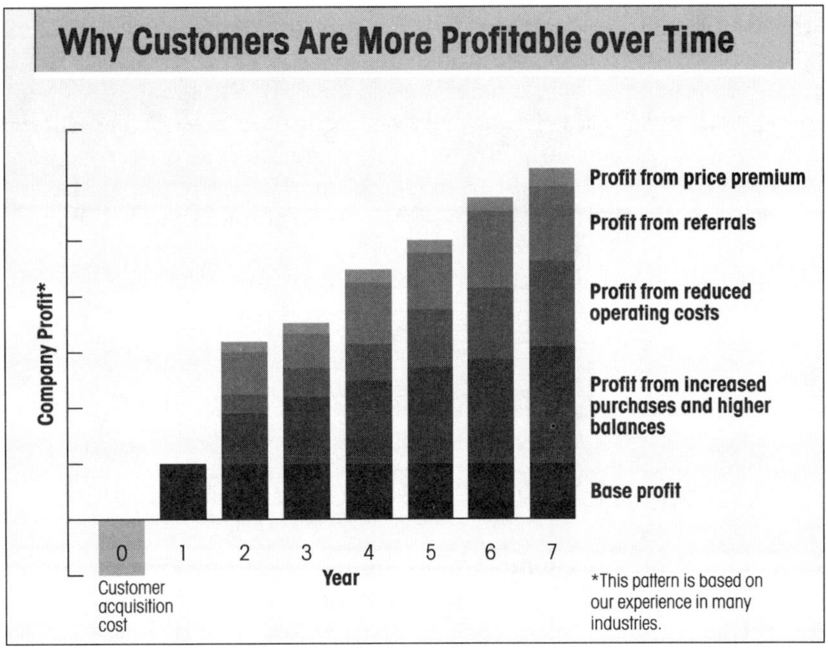

to investments in service quality versus things like cost reduction, for which the objectives have been more tangible.

Knowing that defections are closely linked to profits also helps explain why some companies that have relatively high unit costs can still be quite profitable. Companies with loyal, long-time customers can financially outperform competitors with lower unit costs and high market share but high customer churn. For instance, in the credit card business, a 10% reduction in unit costs is financially equivalent to a 2% decrease in defection rate. Low-defection strategies can overwhelm low-cost strategies.

And understanding the link between defections and profits provides a guide to lucrative growth. It is common for a business to lose 15% to 20% of its customers each year. Simply cutting defections in half will more than double the average company's growth rate. Companies with high retention rates that want to expand through acquisition can create value by acquiring low retention competitors and reducing their defections.

Defections Management

Although service companies probably can't–and shouldn't try to–eliminate all defections, they can and must reduce them. But even to approach zero defections, companies must pursue that goal in a coordinated way. The organization should be prepared to spot customers who leave and then to analyze and act on the information they provide.

Watch the door. Managing for zero defections requires mechanisms to find customers who have ended their relationship with the company–or are about to end it. While compiling this kind of customer data almost always involves the use of information technology of some kind, major investments in new systems are unnecessary.

The more critical issue is whether the business regularly gathers information about customers. Some companies already do. Credit card companies, magazine publishers, direct mailers, life insurers, cellular phone companies, and banks, for example, all collect reams of data as a matter of course. They have at their disposal the names and addresses, purchasing histories, and telephone numbers of all their customers. For these businesses, exposing defections is relatively easy. It's just a matter of organizing the data.

Sometimes, defining a "defection" takes some work. In the railroad business, for instance, few customers stop using your service completely, but a customer that shifts 80% of its shipments to trucks should not be considered "retained." The key is to identify the customer behaviors that both drive your economics and gauge customer loyalty.

For some businesses, the task of spotting defectors is challenging even if they are well defined, because customers tend to be faceless and nameless to management. Businesses like retailing will have to find creative ways to "know" their customers. Consider the example of Staples, the Boston-based office products discounter. It has done a superb job of gathering information usually lost at the cashier or sales clerk. From its opening, it had a database to store and analyze customer information. Whenever a customer goes through the checkout line, the cashier offers him or her a membership card. The card entitles the holder to special promotions and certain discounts. The only requirement for the card is that the person fill out an application form, which asks for things like name, job title, and address. All subsequent purchases are automatically logged against the card number. This way, Staples can accumulate detailed information about buying habits, frequency of visits, average dollar value spent, and particular items purchased.

Even restaurants can collect data. A crab house in Maryland, for instance, started entering into its PC information from the reservation list. Managers can now find out how often particular customers return and can contact those who seem to be losing interest in the restaurant.

What are defectors telling you? One reason to find customers who are leaving is to try to win them back. MBNA America has a customer-defection "swat" team staffed by some of the company's best telemarketers. When customers cancel their credit cards, the swat team tries to convince them to stay. It is successful half of the time.

But the more important motive for finding defectors is for the insight they provide. Customers who leave can provide a view of the business that is unavailable to those on the inside. And whatever caused one individual to defect may cause many others to follow. The idea is to use defections as an early warning signal – to learn from defectors why they left the company and to use that information to improve the business.

Unlike conventional market research, feedback from defecting customers tends to be concrete and specific. It doesn't attempt to measure things like attitudes or satisfaction, which are changeable and subjective, and it doesn't raise hypothetical questions, which may be irrelevant to the respondents. Defections analysis involves specific, relevant questions about why a customer has defected. Customers are usually able to articulate their reasons, and some skillful probing can get at the root cause.

This information is useful in a variety of ways, as the Staples example shows. Staples constantly tracks defections, so when customers stop doing business there or don't buy certain products, the store notices it immediately and calls to get feedback. It may be a clue that the competition is underpricing Staples on certain goods – a competitive factor management can explore further. If it finds sufficient evidence, Staples may cut prices on those items. This information is highly valued because it pinpoints the uncompetitive products and saves the chain from launching expensive broad-brush promotions pitching everything to everybody.

Staples's telemarketers try to discern which merchandise its customers want and don't want and why. The company uses that information to change its buying stock and to target its catalogs and coupons more precisely. Instead of running coupons in the newspaper, for instance, it can insert them in the catalogs it sends to particular customers or industries that have proved responsive to coupons.

Defections analysis can also help companies decide which service-quality investments will be profitable. Should you invest in computerized cash registers or a new phone system? Which of the two will address the most frequent causes of defection? One bank made a large investment to improve the accuracy of monthly account statements. But when the bank began to study defectors, it learned that less than 1% of its customers were leaving because of inaccurate statements.

A company that is losing customers because of long lines can estimate what percentage of defectors it would save by buying new cash registers, and it can use its defection curve to find the dollar value of saving them. Then, using standard investment-analysis techniques, it can compare the cost of the new equipment with the benefit of keeping customers.

Achieving service quality doesn't mean slavishly keeping all customers at any cost. There are some customers the company should not try to serve. If particular types of customers don't stay and become profitable, companies should not invest in attracting them. When a health insurance company realized that certain companies purchase only on the basis of price and switch health insurers every year, for example, it decided not to waste its efforts seeking their business. It told its brokers not to write policies for companies that have switched carriers more than twice in the past five years.

Conversely, much of the information used to find defectors can point to common traits among custom-

A Credit Card Company's Defection Curve

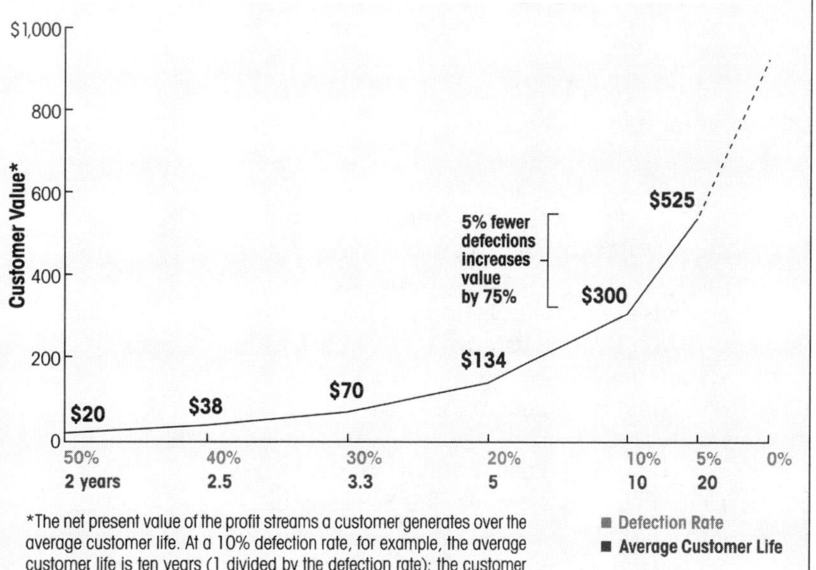

*The net present value of the profit streams a customer generates over the average customer life. At a 10% defection rate, for example, the average customer life is ten years (1 divided by the defection rate); the customer value is the net present value of the profit streams for ten years.

Reducing Defections 5% Boosts Profits 25% to 85%

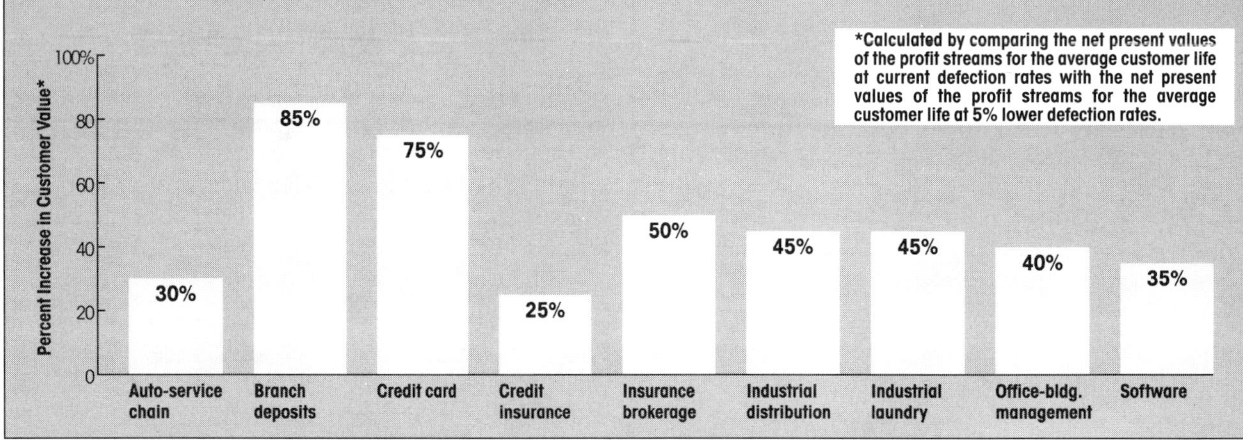

ers who stay longer. The company can use defection rates to clarify the characteristics of the market it wants to pursue and target its advertising and promotions accordingly.

The Zero Defections Culture

Many business leaders have been frustrated by their inability to follow through on their public commitment to service quality. Since defection rates are measurable, they are manageable. Managers can establish meaningful targets and monitor progress. But like any important change, managing for zero defections must have supporters at all organizational levels. Management must develop that support by training the work force and using defections as a primary performance measure.

Everyone in the organization must understand that zero defections is the goal. Mastercare, the auto-service subsidiary of Bridgestone/Firestone, emphasizes the importance of keeping customers by stating it clearly in its mission statement. The statement says, in part, that the company's goal is "to provide the service-buying public with a superior buying experience that will encourage them to return willingly and to share their experience with others." MBNA America sends its paychecks in envelopes labeled "Brought to you by the customer." It also has a customer advocate who sits in on all major decision-making sessions to make sure customers' interests are represented.

It is important to make all employees understand the lifetime value of a customer. Phil Bressler, the co-owner of five Domino's Pizza stores in Montgomery County, Maryland, calculated that regular customers were worth more than $5,000 over the life of a ten-year franchise contract. He made sure that every order taker, delivery person, and store manager knew that number. For him, telling workers that customers were valuable was not nearly as potent as stating the dollar amount: "It's so much more than they think that it really hits home."

Mastercare has redesigned its employee training to emphasize the importance of keeping customers. For example, many customers who stopped doing business with Mastercare mentioned that they didn't like being pressured into repairs they had not planned on. So Mastercare now trains store managers to identify and solve the customer's problem rather than to maximize sales. Videos and role-playing dramatize these different definitions of good service.

Mastercare's message to employees includes a candid admission that previous, well-intentioned incentives had inadvertently caused employees to run the business the wrong way; now it is asking them to change. And it builds credibility among employees by sharing its strategic goals and customer outreach plans. In the two target markets where this approach has been used, results are good. Employees have responded enthusiastically, and 25% more customers say they intend to return.

Senior executives at MBNA America learn from defecting customers. Each one spends four hours a month in a special "listening room" monitoring routine customer service calls as well as calls from customers who are canceling their credit cards.

> Great-West Life Assurance pays brokers a premium for lowering customer defection rates.

Beyond conveying a sense of urgency, training should teach employees the specifics of defections analysis, like how to gather the information, whom to pass it on to, and what actions to take in response. In one company's branch banking system, retention data is sent monthly to the regional vice presidents and branch managers for review. It allows the regional vice presidents to identify and focus on branches that most need to improve service quality, and it gives branch managers quick feedback on performance.

Employees will be more motivated if incentives are tied to defection rates. MBNA, for example, has determined for each department the one or two things that have the biggest impact on keeping customers. Each department is measured daily on how well performance targets are met. Every morning, the previous day's performance is posted in several places throughout the building. Each day that the company hits 95% of these performance targets, MBNA contributes money to a bonus pool. Managers use the pool to pay yearly bonuses of up to 20% of a person's salary. The president visits departments that fall short of their targets to find out where the problem lies.

Great-West Life Assurance Company of Englewood, Colorado also uses incentives effectively. It pays a 50% premium to group-health-insurance brokers that hit customer-retention targets. This system gives brokers the incentive to look for customers who will stay with the company for a long time.

Having everyone in the company work toward keeping customers and basing rewards on how well they do creates a positive company atmosphere. Encouraging employees to solve customer problems and eliminate the source of complaints allows them to be "nice," and customers treat them better in return. The overall exchange is more rewarding, and people enjoy their work more. Not just customers but also employees will want to continue their relationship with the business. MBNA is besieged by applicants for job openings, while a competitor a few miles away is moving some of its operations out of the state because it can't find enough employees.

The success of MBNA shows that it is possible to achieve big improvements in both service quality and profits in a reasonably short time. But it also shows that focusing on keeping customers instead of simply having lots of them takes effort. A company can leverage business performance and profits through customer defections only when the notion permeates corporate life and when all organizational levels understand the concept of zero defections and know how to act on it.

> Employees like to work for a company that keeps its customers. MBNA America has far more applicants than jobs.

Trying to retain all of your profitable customers is elementary. Managing toward zero defections is revolutionary. It requires careful definition of defection, information systems that can measure results over time in comparison with competitors, and a clear understanding of the microeconomics of defection.

Ultimately, defections should be a key performance measure for senior management and a fundamental component of incentive systems. Managers should know the company's defection rate, what happens to profits when the rate moves up or down, and why defections occur. They should make sure the entire organization understands the importance of keeping customers and encourage employees to pursue zero defections by tying incentives, planning, and budgeting to defection targets. Most important, managers should use defections as a vehicle for continuously improving the quality and value of the services they provide to customers.

Just as the quality revolution in manufacturing had a profound impact on the competitiveness of companies, the quality revolution in services will create a new set of winners and losers. The winners will be those who lead the way in managing toward zero defections.

Reprint 90508

How the best companies turn complaining customers into loyal ones.

The Profitable Art of Service Recovery

by Christopher W.L. Hart, James L. Heskett, and W. Earl Sasser, Jr.

Mistakes are a critical part of every service. Hard as they try, even the best service companies can't prevent the occasional late flight, burned steak, or missed delivery. The fact is, in services, often performed in the customer's presence, errors are inevitable.

But dissatisfied customers are not. While companies may not be able to prevent all problems, they can learn to recover from them. A good recovery can turn angry, frustrated customers into loyal ones. It can, in fact, create more goodwill than if things had gone smoothly in the first place. Consider how Club Med-Cancun, part of the Paris-based Club Mediterranée, recovered from a service nightmare and won the loyalty of one group of vacationers.

The right response to mistaken billings and late deliveries can cement customer loyalty.

The vacationers had nothing but trouble getting from New York to their Mexican destination. The flight took off 6 hours late, made 2 unexpected stops, and circled for 30 minutes before it could land. Because of all the delays and mishaps, the plane was en route for 10 hours more than planned and ran out of food and drinks. It finally arrived at 2 o'clock in the morning, with a landing so rough that oxygen masks and luggage dropped from overhead. By the time the plane pulled up to the gate, the soured passengers were faint with hunger and convinced that their vacation was ruined before it had even started. One lawyer on board was already collecting names and addresses for a class-action lawsuit.

Silvio de Bortoli, the general manager of the Cancun resort and a legend throughout the organization for his ability to satisfy customers, got word of the horrendous flight and immediately created an antidote. He took half the staff to the airport, where they laid out a table of snacks and drinks and set up a stereo system to play lively music. As the guests filed through the gate, they received personal greetings, help with their bags, a sympathetic ear, and a chauf-

Christopher W.L. Hart is president of the TQM Group in Cambridge, Massachusetts. James L. Heskett and W. Earl Sasser, Jr. are professors of production and operations management at the Harvard Business School. Together they wrote Service Breakthroughs: Changing the Rules of the Game, *which Free Press will publish this fall.*

feured ride to the resort. Waiting for them at Club Med was a lavish banquet, complete with mariachi band and champagne. Moreover, the staff had rallied other guests to wait up and greet the newcomers, and the partying continued until sunrise. Many guests said it was the most fun they'd had since college.

In the end, the vacationers had a better experience than if their flight from New York had gone like clockwork. Although the company probably couldn't measure it, Club Mediterranée won market share that night. After all, the battle for market share is won not by analyzing demographic trends, ratings points, and other global measures but rather by pleasing customers one at a time.

Opportunities for service recovery abound. Any problem that employees who are close to the customer can discover and resolve is a chance to go beyond the call of duty and win a customer for life. We're not talking about gas leaks in Bhopal or Tylenol poisonings, which threaten large-scale damage and demand top management's attention. We're talking about mistaken billings and late deliveries, the seemingly small issues that can ignite a person's temper. The stuff angry letters to the chief executive are made of.

It's tempting to dismiss the occasional problem as petty and complaining customers as cranks, but managers should resist those easy outs. No business can afford to lose customers, if only because it costs much more to replace a customer than it does to retain one–five times more, most industry experts agree. Companies that alienate and frustrate their customers will soon have none left to bother them. Those that go out of their way to please customers will soon have many more.

Good recoveries from service problems do happen, but usually because some exceptional individual like de Bortoli takes the initiative to solve a customer's problem. Companies should not depend on such rare instances of resourcefulness. They should take steps to ensure that everyone in the organization has the skill, motivation, and authority to make service recovery an integral part of operations.

The Road to Service Recovery

Service companies must become gymnasts, able to regain their balance instantly after a slipup and continue their routines. Such grace is earned by focusing on the goal of customer satisfaction, adopting a customer-focused attitude, and cultivating the special skills necessary to recovery.

Ironically, recovery skills come especially hard to companies that joined the quality-control movement and have spent the past decade making their service-delivery systems streamlined and efficient. Accepting admonishments to "do the right thing" and adopting manufacturing's philosophy of "zero defects," they developed rigid systems to achieve it. They introduced sophisticated technologies and enacted strict policies to control employee behavior. The idea was to ensure that even uneducated, unmotivated workers could consistently deliver high-quality service. By the 1980s, many such systems had

Sometimes it takes more than an apology to make amends.

been developed–to improve service in everything from scheduling airline departures and posting banking transactions to maintaining hotel rooms. All of these systems were modeled after assembly-line production systems.

These production-oriented service-delivery systems have gone a long way toward achieving consistently high service standards. But they're not bulletproof. The fact is that in services, no matter how rigorous the procedures and employee training or how advanced the technology, zero defects is an unattainable goal. Unlike manufacturers that can adjust the inputs and machinery until products are uniformly perfect, service companies cannot escape variation. Factors like the weather and the customers themselves are beyond a company's control. The best airline reservation system can't prevent the airport from fogging over. A restaurant that creates the most artistic food presentations can't prevent a customer from disliking the taste.

> Companies have made service delivery idiot-proof. But idiots can't solve problems.

When the inevitable problems arise, customers are almost always disappointed. The typical service delivery system is completely unprepared to deal with exceptions. Studies we've done show that more than half of all efforts to respond to customer complaints actually *reinforce* negative reactions to a service.

The surest way to recover from service mishaps is for workers on the front line to identify and solve the customer's problem. Doing so requires decision making and rule breaking–exactly what employees have been conditioned against. Workers have been taught that it's not their job to alter the routine. Even if they'd like to help the customer, they are frustrated by the fact that they are not allowed to do it. Worse yet, they don't know how. We have all heard the typical responses: "It's not my fault." "It's not in the computer." "I'll have to ask my supervisor." Meanwhile, airline passengers stew over the meetings they've missed; restaurant patrons return home hungry and annoyed.

Companies shouldn't abandon their production-oriented systems, but they should complement them with an equal facility for service recovery. They should be as comfortable with the exceptions as they are with the rules. Developing the perspective to recognize service-recovery opportunities and the skills to act on them is clearly an effort, but one well worth making.

Mistreating customers can have a devastating effect on the business, as this example shows. John Barrier, a 30-year customer of a bank in Spokane, Washington, parked his car in a lot owned by the bank while he did business across the street. An attendant told him he could get the parking validated if he did business at the bank, which was not his usual branch. Barrier cashed a check, but afterward was refused the validation because he had not made a deposit. He patiently explained to the receptionist that he was a long-time customer and that he had millions of dollars in the bank's checking, investment, and trust accounts. She was unmoved. He less patiently explained the situation to the branch manager, to no avail.

The customer paid for the parking, but he was so steamed that he drove 40 blocks to his home branch and explained the incident to his usual banker. He said if he didn't receive a phone call by the end of the day, he would close all his accounts. The call never came, he made his first withdrawal of $1 million, and Americans across the country heard the story when it hit the evening news. Needless to say, the bank executives were embarrassed and worked hard to persuade the customer to give the bank another chance.

Companies that want to build the capability of recovering from service problems should do these things: measure the costs of effective service recovery, break customer silence and listen closely for complaints, anticipate needs for recovery, act fast, train employees, empower the front line, and close the customer feedback loop.

Measure the Costs

Measurement precedes management. This is especially true for service recovery–managers often underestimate the profits lost when a customer departs unhappy, and therefore they undermanage ways of avoiding such losses. They concentrate on attracting new customers that may actually represent unprofitable business and neglect to take steps to retain more valuable existing customers. Measurement often is the only way of getting top management's attention. What gets measured is truly what gets managed here.

Errors have certain costs associated with them. Some take the form of money-back guarantees, warranty work, or replacements, which fall on the company. But dissatisfied customers almost always get

1. U.S. Office of Consumer Affairs, *Consumer Complaint Handling in America: An Update Study, Part II* (Washington, D.C.: Technical Assistance Research Programs Institute, April 1, 1986), p. 50.

stuck with certain costs – the money they spend for phone calls, the time they spend making their cases, and the aggravation they must endure throughout. The customer left stranded on the highway because her car was not repaired properly might miss an important meeting, have to pay for a tow truck, and spend time waiting for the repair to be made. Many service companies conveniently overlook these hidden costs, but the customer surely won't. Companies known for excellent service will go the extra yard to cover all the costs a failure incurs or, if the inconvenience is so great that the company cannot completely compensate the customer, the tone of the response must signal the company's regret.

A study done for the U.S. Office of Consumer Affairs found that in households with service problems with potential costs of more than $100, 54% would maintain brand loyalty if their problems were satisfactorily resolved. Only 19% would repeat their purchase if they were unhappy with the problem resolution. For less expensive problems ($1 to $5), 70% would maintain brand loyalty if their problems were resolved satisfactorily; only 46% would repurchase if the problem wasn't fixed.[1]

Maine Savings Bank pays customers $1 for each suggestion letter.

Considering how much it costs to lose a customer, few recovery efforts are too extreme. At Club Med, one lost customer costs the company at least $2,400: a loyal guest visits the resorts an average of four times after the initial visit and spends roughly $1,000 each time. The contribution margin is 60%. So when a Club Med customer doesn't return, the company loses 60% of $4,000, or $2,400. It also has to replace that customer through expensive marketing efforts.

Break the Silence

Every customer's problem is an opportunity for the company to prove its commitment to service – even if the company is not to blame. The theatergoer who forgets his ticket will long be grateful if the usher slips him in. The service experiences customers rave about most are those in which they were at fault but the company responded anyway.

Of course, you can't solve a customer's problem until you know what it is, so all good recoveries start with identifying the sore spots. Service company executives know very well that some customers make a point of being heard. They write letters, make phone calls, ask to speak to the manager – and the manager's manager. Listening to these customers is important.

But companies shouldn't attend to just the squeaky wheels. They must also be active problem finders. Complainers are the exception; most unhappy people don't speak up. They may think the situation is hopeless, or they don't want to create a scene. Or they can't be bothered to write a letter or make a phone call – it's one more hassle in what may have been a string of many. There are many ways service businesses can encourage this "silent majority" of dissatisfied customers to identify themselves so the company can win them back.

The simplest way is to make it easy for customers to complain. Many businesses have established "800" numbers so customers can report problems easily and at the company's expense. American Express has installed such lines and estimates that it achieves responses more quickly and at 10% to 20% of the cost of handling correspondence. Marriott Corporation has a 24-hour "hot line" in its hotels, which makes it easy for guests to complain on the spot. It's worth adding that hot lines should be staffed sufficiently to prevent a common shortcoming of centralized telephone systems – too few incoming lines. Repeated busy signals do nothing to improve customers' attitudes.

A more direct way to solicit complaints is to ask a simple question like, "How was everything?" Many customers who can't be bothered filling out forms or making calls will volunteer their impressions of the service they received when asked to. And even if the customer has nothing to say, the company has sent the signal that it cares. That's why British Airways has installed what it calls Video Point booths at Heathrow Airport in London so that travelers can tape their reactions upon arrival. Customer service representatives view the tapes and respond. Maine Savings Bank in Portland offers its patrons $1 for every letter they write suggesting ways to improve service. The bank averages more than 500 letters a year from customers who might have kept their ideas to themselves; it extended its lobby hours after many customers made that suggestion.

> Solve the customer's problem – even if it isn't the company's fault.

An even more aggressive approach to unearthing problems is to look for trouble in the making – to listen carefully for offhand comments customers make and to tune into and anticipate their needs. An alert employee of a Marriott hotel that caters to businesspeople overheard a guest fretting about the lack of privacy in the concierge lounge where he wanted to hold an impromptu meeting with a few colleagues. The hotel worker called the front desk and arranged for a vacant suite so the guest could hold his meeting without distractions.

A participant in a service-recovery workshop shared this striking example of a company that recognized a problem it didn't create but nonetheless was prepared to solve:

"A while back, there was a very bad fire in my house. The next day I was raking through my possessions, my family sitting on the front stoop, when a Domino's Pizza truck pulled up. The driver got out and approached us with two pizzas. I told him I didn't order any pizza and explained that our house had just burned. 'I know,' he replied. 'I saw you when I drove by half an hour ago. I figured you must be really hungry, so my store manager and I decided to make a couple of pizzas for you. We put everything on them. If that's not how you like them, I'll take them back and get them made the way you like – on the house.'

"I couldn't believe it," the participant concluded. "Do you think I'd ever buy pizza from anyone else?"

More formal "listening devices" like questionnaires and customer suggestion boxes are effective only if someone monitors them continually and acts on complaints and suggestions in a timely fashion.

Stew Leonard's, a retail dairy store in Norwalk, Connecticut renowned for its service, has a suggestion box, which pays off handsomely. One evening at about 6 P.M., Stew Leonard, Jr. found a crisply worded complaint written by a customer just a half-hour earlier. "I made a special stop on my way home from work to buy chicken breasts for dinner, but you're sold out. Now I'll have to eat a TV dinner instead." As Leonard was reading the complaint, a Perdue chicken truck happened to pull up to the store's loading dock. Within minutes, someone was heading off to the frustrated customer's house with a complimentary two-pound package of fresh chicken breasts.

Anticipate Needs for Recovery

Companies can narrow the search for problems (read "opportunities") by monitoring certain areas of the organization and addressing them in their service-recovery strategies. Complex scheduling that involves coordinating the movement of people or equipment, for instance, tends to be problem-prone. There, one error can trigger a devastating chain reaction, as in airline flight cancellations.

We were among a group of professors studying service excellence that recently felt the brunt of a failure to anticipate the need for recovery. En route from Boston to Columbus, Ohio with a stop at Washington National Airport, the group was not warned that, because of a take-off curfew at Washington National, a late departure from Boston might mean the flight would not be allowed to continue to Ohio. In fact, the US Air plane was held overnight in Washington, making it impossible for the group to make it to its meeting the next morning. Worse, the harried late-night staff at US Air counters was shorthanded, so already disgruntled passengers had to stand in line for an hour for assistance. If ever there was a problem waiting to happen, this was it. Because the company failed to anticipate it, both US Air's frontline employees and its customers suffered.

New services and products also tend to create confusion and spawn unexpected requests. When Dallas-Fort Worth International Airport first opened, it proved very confusing to passengers. Not only was the airport of unusual design and size but it also included one of the first rail systems to shuttle passengers between terminals. It took the airport months to change the layout and reposition signs to make the airport more "user friendly"–months of confusion and frustration for those who used it.

Areas where turnover is high and workers are therefore inexperienced are also worth watching. The first person most air travelers encounter is the operator of the security gate. Of all airline employees travelers meet during a trip, this person is probably the lowest paid and the least experienced in handling people–yet he or she has the job of creating a positive first impression. Security is therefore a leading candidate for preparation for airline service recovery.

Fixing customers' problems as they crop up is good and is necessary. But service companies should put those isolated incidents to use by tracking problems to see where they occur most often and which ones tend to recur. They can then prepare the organization. When the CEO of an insurance company locked the keys in the trunk of his rental car at a Sheraton Hotel in Boca Raton, Florida, the staff rose to the occasion. The bellman informed the CEO that a locksmith under contract with the hotel would replace the keys within 15 minutes. Meanwhile, other staff members used a rolling auto-jack stored nearby to jack up the car and push it out of the driveway where it was blocking check-in traffic. Obviously, it wasn't the first time someone had lost a set of car keys.

Act Fast

Identifying a problem quickly–even before it registers with the customer–is fruitful only if the company responds fast. Our most recent research suggests that customers who have bad experiences tell approximately 11 people about it; those with good experiences tell just 6.

Service problems quickly escalate, so the opportunity to prove one's commitment to the customer is fleeting, especially if the company is at fault. In general, the company's first priority should be to complete the service promptly. The customer whose car breaks down because it was serviced improperly wants her car fixed. The bank customer whose account is in error wants it properly credited.

Paul Hawken, CEO of Smith & Hawken, a garden supply mail-order company based in Mill Valley, California, found a simple way to speed response time: by using the phone instead of mail. Early in the company's history, customers with routine inquiries and problems received form letters in reply. Management found that approach to be cost-effective, but

Smith & Hawken customers get phone calls–not form letters–when problems arise.

the savings came at the customer's expense. In one instance, an old and valuable customer placed an order on an American Express card, which the company mistakenly thought was invalid. It sent the usual form letter. The customer wrote back. The company sent another form letter. The customer wrote back. When a manager finally realized that the order had been held up two months because of crossed correspondence, he was horrified. The company prided itself on customer service, yet it had managed to anger and frustrate one of its best clients.

The person who made the discovery immediately sent the woman her order and swallowed the $90 charge. And the company learned a lesson: from that moment on, Smith & Hawken stopped using the mail to handle questions or problems: now someone picks up the phone and resolves them in minutes instead of months. And the cost of the phone call is offset by the elimination of paperwork.[2]

The urgent resumption of service and an apology are often sufficient to make amends. But not always. Some situations call for a gesture that clearly says, "We realize there's been a mistake, and we want to make it up to you." Many restaurants will automatically give patrons a glass of wine or a free dessert if the wait for a table is too long. First Union National Bank in Charlotte, North Carolina sends a dozen roses to customers who have been badly inconvenienced. In some cases, branch managers or bank executives personally deliver the flowers. Such extraordinary efforts require extraordinary preparation at every organizational level.

Train Employees

The organization must train the people who interact directly with customers, and then it must empower them. That is, it must give them the authority, responsibility, and incentives to recognize, care about, and attend to customer needs. Empowering the bottom of the organizational pyramid can be threatening, especially to middle-level managers, who may read it as an erosion of their own authority and worth. But it is absolutely essential to good service recovery. Employees close to the customer are the first to know about problems and are in the best position to determine what can be done to satisfy the customer.

Training can go a long way in developing the communications skills and creative thinking needed to deal with irate customers. Recovery training should focus on teaching employees how to make decisions on their feet and on developing an awareness of customers' concerns.

The most effective way to develop recovery skills is through simulated real-life situations and role playing. Agents for the U.S. Secret Service, who are among the world's most highly trained recovery specialists, go through a comprehensive assortment of recovery drills before they work in the field. Agents spend hours thinking about what might happen and discussing and planning for all the possible contingencies. Each agent – no matter how senior – goes through follow-up training sessions once a year.

Sonesta Hotels uses games as part of its orientation program for new employees. Trainees are divided into two teams, each of which in turn receives a description of a problem and is asked to come up with a solution. The opposing team has five possible answers to compare the response with. Points are awarded depending on how well the responses fit the general criteria of keen observation, responsiveness, care and concern, and compensation for true loss.

The American Association of Homes for the Aging distributes a board game, "The Game of Aging Concerns," which retirement homes use to give workers practice in dealing with tough issues. The game consists of a board, six playing pieces, a die, and 40 cards, each describing a real problem or role-playing situation. One card reads: "Mr. Talbot, 78, complains, 'Your facility is too cold and makes me ill.' He demands that you turn on the heat. Other residents want the air-conditioning left on. The heating/air-conditioning system is centrally controlled." Another reads, "A 72-year-old man who is intoxicated walks into the dining room and acts disruptive."

As players draw cards, they read them aloud, describe how they feel, and say what they would do. The idea is to expose players to a variety of concepts, give them feedback and reinforcement, and allow them to internalize a set of criteria for evaluating real-life problems. And the spirit of the game paves the way for an ongoing dialogue among workers.

To help develop recovery skills, training should also give people a sense of the whole organization. Extreme specialization of tasks gives operators tunnel vision, which makes it hard to see problems in the making. A worker who understands the entire service delivery process is more likely to understand the interconnectedness of the system and find a quick solution. The most direct way to develop this perspective is by rotating workers through different jobs and departments.

Managers know how to solve customers' problems, but people don't want to wait for their concerns

2. From Paul Hawken, *Growing a Business* (New York: Simon & Schuster, 1988).

to travel through the organizational hierarchy or to bounce from one department to another. Nor can managers answer every phone, stand behind every counter, process every piece of paper. So service recovery ultimately rests on the shoulders of employees on the front line. This implies a very different role for employees who have direct customer contact. In addition to following rules, sticking to a routine, and treating every situation alike no matter what, front-line workers must be able to do the opposite: bend the rules, take initiative, and improvise. Building a staff that can do both requires rigorous and conscious effort and is at the heart of a company's ability to recover from service mishaps.

Empower the Front Line

Training can give employees the perspective that service recovery requires, but the company must empower them to act. It must give employees the authority, responsibility, and incentives to follow through with customers.

The authority to act refers to the set of resources the employee has access to and the decisions they are permitted to make. In most service companies, only managers can spend money or otherwise make things happen. Organizations that empower workers make it clear that they are permitted to use their judgment to make phone calls, credit accounts, or send flowers. For example, Montgomery Ward chairman Bernard F. Brennan has authorized its 7,700 salesclerks to approve checks and handle merchandise-return problems, functions that once were reserved for store managers.

Responsibility goes one step further. It says that employees are *supposed* to recognize and attend to customers' needs. Managers can instill that sense of purpose in many ways—through training and reminders in newsletters and by developing an overall environment that puts customers first.

Responsibility means the obligation to act, not just to accept blame. One couple was visiting a resort on the coast of Mexico in the middle of August when the water main from the town broke. The heat and humidity were ferocious, and the hotel was without air-conditioning and water for most of four days. Repair attempts were feeble. The water would run for just a few hours before things would fail again. Each time the system broke, the general manager gathered the guests in the lounge and prattled on about how he took "full responsibility" for everything. When he said *responsibility* he meant blame. What customers wanted, of course, was action.

Good service companies rely on "standard operating procedures" for problems that come up from time to time. At McDonald's, for instance, employees know that when a customer complains that his burger is cold, they should automatically give him a fresh one, no questions asked. For other situations, management can only establish guidelines. Many of the problems at the Minneapolis Marriott City Center are one-of-a-kinds, so management has authorized employees to spend $10 at their discretion to satisfy guests. Once when a guest complained mildly about not being able to find a particular book in the hotel gift shop, the cashier, at the end of her shift, walked to a local bookstore, purchased the book with her $10, and delivered it to the guest's room. The guest was, of course, astonished.

The Minneapolis Marriott also put together a "Sweet Dreams" package consisting of a cordial, a small bud vase with a carnation, and homemade cookies. Hotel staff members are encouraged to give it to customers who are having difficulties the hotel can't otherwise fix. When one guest mentioned that her flight was four hours late and that she was exhausted, the staff person at the front desk sent a "Sweet Dreams" package to her room. Another associate noticed a guest with a hacking cough; she included a box of cough drops with the package.

A company's reward structure must give employees positive reinforcement for solving problems and pleasing customers—not just for reducing the number of complaints. Good recoveries should be publicized and held up as examples to inspire others. Federal Express created the "Golden Falcon" and "Bravo Zulu" awards to recognize service and recovery excellence. Winners get a gold pin, recognition in the company newsletter, and a phone call from the chief operating officer as well as ten shares of company stock.

As the nature of jobs with high customer contact changes, companies may find it necessary to change their hiring requirements or reassign workers. Embracing the notion of empowerment, the Marriott Desert Springs Resort revised the job description for its front-desk clerk/cashiers, room controllers, and front-desk supervisors: the major—indeed the *only*—goal of these positions is to ensure that "our guests experience excellent service and hospitality while staying at our resort." The resort charges the people

> Think about the *value* of pleasing a customer—not the cost.

in those positions with learning the correct technical procedures, using their authority to do anything to keep guests happy, using their power to satisfy guests on the spot without hassle, assisting in finding the ultimate cause for guests' problems, and informing managers of ways to improve the overall hotel, working conditions, or guests' comfort.

If this seems logical, why don't more services do it? Many are afraid that empowered employees will "give away the store." They overlook the fact that tying employees' hands effectively guarantees that some customers won't return. The cost of keeping those customers is small compared with what it takes to replace them.

Close the Loop

If a customer's complaint leads to corrective measures, the company should tell the customer about the improvement. Closing the loop makes customers feel as if they're part of an extended quality-control team. If it's something that can't be fixed, the company should simply explain why.

In our service-management classes, we ask each student to send one letter of commendation and one letter of criticism to service companies where they had memorable experiences. All contain suggestions for service improvements. The responses to these hundreds of letters from the real world are revealing. Only about two-thirds of the companies that receive thoughtful, constructive, but critical letters bother to respond. A smaller proportion of the companies receiving praise write back. Worse yet, more than half of all responses to letters of criticism either reinforce or fail to counter the senders' original negative perceptions.

Effective ways of closing the loop include making timely telephone responses, asking the customer for even more feedback, and letting the customer know how his or her suggestions might be implemented. These efforts tend to give customers a more positive impression than various forms of remuneration.

Brilliant Recoveries

Recovery is a different management philosophy, one that embraces customer satisfaction as a primary goal of business. This mind-set can change the rules of the game for service companies. It shifts the emphasis from the *cost* of pleasing a customer to the *value* of doing so, and it entrusts frontline employees with using their judgment.

While many organizations pay lip service to the notion of serving customers, few are wholly committed to it. Even those that see the connection between 100% customer satisfaction and the ultimate success of the business are often ill-equipped to retain dissatisfied patrons.

But recovery is fundamental to service excellence and should therefore be regarded as an integral part of a service company's strategy. When a worker at a San Antonio hotel accidentally broke the glass vase a guest had carried from Mexico, the resident manager drove 150 miles to replace it. Customers remember such experiences. In service businesses, the old adage must be revised: To err is human; to recover, divine.

Reprint 90407

No-quibble guarantees are self-fulfilling – they promise quality and produce it.

The Power of Unconditional Service Guarantees

by CHRISTOPHER W.L. HART

When you buy a car, a camera, or a toaster oven, you receive a warranty, a guarantee that the product will work. How often do you receive a warranty for auto repair, wedding photography, or a catered dinner? Virtually never. Yet it is here, in buying services, that the assurance of a guarantee would presumably count most.

Many business executives believe that, by definition, services simply can't be guaranteed. Services are generally delivered by human beings, who are known to be less predictable than machines, and they are usually produced at the same time they are consumed. It is one thing to guarantee a camera, which can be inspected before a customer sets eyes on it and which can be returned to the factory for repairs. But how can you preinspect a car tune-up or send an unsuccessful legal argument or bad haircut back for repair? Obviously you can't.

Al Burger *started* with an unconditional guarantee and built his company around it.

But that doesn't mean customer satisfaction can't be guaranteed. Consider the guarantee offered by "Bugs" Burger Bug Killers (BBBK), a Miami-based pest-extermination company that is owned by S.C. Johnson & Son.

Most of BBBK's competitors claim that they will reduce pests to "acceptable levels"; BBBK promises to eliminate them entirely. Its service guarantee to hotel and restaurant clients promises:

- You don't owe one penny until all pests on your premises have been eradicated.
- If you are ever dissatisfied with BBBK's service, you will receive a refund for up to 12 months of the company's services – plus fees for another exterminator of your choice for the next year.
- If a guest spots a pest on your premises, BBBK will pay for the guest's meal or room, send a letter of apology, and pay for a future meal or stay.
- If your facility is closed down due to the presence of roaches or rodents, BBBK will pay any fines, as well as all lost profits, *plus* $5,000.

In short, BBBK says, "If we don't satisfy you 100%, we don't take your money."

Christopher W.L. Hart is an assistant professor at the Harvard Business School, where he teaches a course on service management. As a researcher and consultant, he helps companies design and implement service-guarantee and quality-improvement programs.

How successful is this guarantee? The company, which operates throughout the United States, charges up to ten times more than its competitors and yet has a disproportionately high market share in its operating areas. Its service quality is so outstanding that the company rarely needs to make good on its guarantee (in 1986 it paid out only $120,000 on sales of $33 million–just enough to prove that its promises aren't empty ones).

A main reason that the "Bugs" Burger guarantee is a strong model for the service industry is that its founder, Al Burger, began with the concept of the unconditional guarantee and worked backward, designing his entire organization to support the no-pests guarantee–in short, he started with a vision of error-free service. In this article, I will explain why the service guarantee can help your organization institutionalize superlative performance.

What a Good Service Guarantee Is

Would you be willing to offer a guarantee of 100% customer satisfaction–to pay your dissatisfied customer to use a competitor's service, for example? Or do you believe that promising error-free service is a crazy idea?

Not only is it not crazy, but *committing* to error-free service can help force a company to *provide* it. It's a little like skiing. You've got to lean over your skis as you go down the hill, as if willing yourself to fall. But if you edge properly, you don't fall or plunge wildly; you gain control while you pick up speed.

Similarly, a strong service guarantee that puts the customer first doesn't necessarily lead to chaos and failure. If designed and implemented properly, it enables you to get control over your organization–with clear goals and an information network that gives you the data you need to improve performance. BBBK and other service companies show that a service guarantee is not only possible–it's a boon to performance and profits and can be a vehicle to market dominance.

Most existing service guarantees don't really do the job: they are limited in scope and difficult to use. Lufthansa guarantees that its customers will make their connecting flights *if* there are no delays due to weather or air-traffic control problems. Yet these two factors cause fully 95% of all flight delays. Bank of America will refund up to six months of checking-account fees if a customer is dissatisfied with any aspect of its checking-account service. However, the customer must close the account to collect the modest $5 or $6 per month fee. This guarantee won't win any prizes for fostering repeat business–a primary objective of a good guarantee.

A service guarantee loses power in direct proportion to the number of conditions it contains. How effective is a restaurant's guarantee of prompt service *except* when it's busy? A housing inspector's guarantee to identify all potential problems in a house *except for* those not readily apparent? Squaw Valley in California guarantees "your money back" to any skier who has to wait more than ten minutes in a lift line. But it's not that easy: the skier must first pay $1 and register at the lodge as a beginner, intermediate, or expert; the guarantee is operative only if *all* lifts at the skier's skill level exceed the ten minutes in any half-hour period; and skiers must check with a "ski hostess" at the end of the day to "win" a refund. A Squaw Valley spokesperson said the resort had made just one payout under the guarantee in a year and a half. No wonder!

What is a good service guarantee? It is (1) unconditional, (2) easy to understand and communicate, (3) meaningful, (4) easy (and painless) to invoke, and (5) easy and quick to collect on.

Unconditional. The best service guarantee promises customer satisfaction unconditionally, without exceptions. Like that of L.L. Bean, the Freeport, Maine retail store and mail-order house: "100% satisfaction in every way…." An L.L. Bean customer can return a product at any time and get, at his or her option, a replacement, a refund, or a credit. Reputedly, if a customer returns a pair of L.L. Bean boots after ten years, the company will replace them with new boots and no questions. Talk about customer assurance!

DRAWINGS BY PAUL MEISEL

Customers shouldn't need a lawyer to explain the "ifs, ands, and buts" of a guarantee–because ideally there shouldn't be any conditions; a customer is either satisfied or not.

If a company cannot guarantee all elements of its service unconditionally, it should unconditionally guarantee the elements that it can control. Lufthansa cannot promise on-time arrival, for example, but it could guarantee that passengers will be satisfied with its airport waiting areas, its service on the ground and in the air, and its food quality–or simply guarantee overall satisfaction.

Easy to Understand and Communicate. A guarantee should be written in simple, concise language that pinpoints the promise. Customers then know precisely what they can expect and employees know precisely what's expected of them. "Five-minute" lunch service, rather than "prompt" service, creates clear expectations, as does "no pests," rather than "pest control."

Meaningful. A good service guarantee is meaningful in two respects. First, it guarantees those aspects of your service that are important to your customers. It may be speedy delivery. Bennigan's, a restaurant chain, promises 15-minute service (or you get a free meal) at lunch, when many customers are in a hurry to get back to the office, but not at dinner, when fast service is not considered a priority to most patrons.

In other cases, price may be the most important element, especially with relatively undifferentiated commodities like rental cars or commercial air travel. By promising the lowest prices in town, stereo shops assuage customers' fears that if they don't go to every outlet in the area they'll pay more than they ought to.

L.L. Bean will replace its boots–even after ten years' use.

Second, a good guarantee is meaningful financially; it calls for a significant payout when the promise is not kept. What should it be–a full refund? An offer of free service the next time? A trip to Monte Carlo? The answer depends on factors like the cost of the service, the seriousness of the failure, and customers' perception of what's fair. A money-back payout should be large enough to give customers an incentive to invoke the guarantee if dissatisfied. The adage "Let the punishment fit the crime" is an appropriate guide. At one point, Domino's Pizza (which is based in Ann Arbor, Michigan but operates worldwide) promised "delivery within 30 minutes or the pizza is free." Management found that customers considered this too generous; they felt uncomfortable accepting a free pizza for a mere 5- or 15-minute delay and didn't always take advantage of the guarantee. Consequently, Domino's adjusted its guarantee to "delivery within 30 minutes or $3 off," and customers appear to consider this commitment reasonable.

Easy to Invoke. A customer who is already dissatisfied should not have to jump through hoops to invoke a guarantee; the dissatisfaction is only exacerbated when the customer has to talk to three different people, fill out five forms, go to a different location, make two telephone calls, send in written proof of purchase with a full description of the events, wait for a written reply, go somewhere else to see someone to verify all the preceding facts, and so on.

Traveler's Advantage–a division of CUC International–has, in principle, a great idea: to guarantee the lowest price on the accommodations it books. But to invoke the guarantee, customers must prove the lower competing price by booking with another agency. That's unpleasant work. Cititravel, a subsidiary of Citicorp, has a better approach. A customer who knows of a lower price can call a toll-free number and speak with an agent, as I did recently. The agent told me that if I didn't have proof of the lower fare, she'd check competing airfares on her computer screen. If the lower fare was there, I'd get that price. If not, she would call the competing airline. If the price was confirmed, she said, "We'll refund your money so fast, you won't believe it–because we want you to be our customer." That's the right attitude if you're offering a guarantee.

Similarly, customers should not be made to feel guilty about invoking the guarantee–no questioning, no raised eyebrows, or "Why me, Lord?" looks. A company should encourage unhappy customers to invoke its guarantee, not put up roadblocks to keep them from speaking up.

Easy to Collect. Customers shouldn't have to work hard to collect a payout, either. The procedure should be easy and, equally important, quick–on the spot, if possible. Dissatisfaction with a Manpower temporary worker, for instance, results in an immediate credit to your bill.

What you should *not* do in your guarantee: don't promise something your customers already expect; don't shroud a guarantee in so many conditions that it loses its point; and don't offer a guarantee so mild

1. See British Airways study cited in Karl Albrecht and Ron Zemke, *Service America!* (Homewood, Ill.: Dow Jones-Irwin, 1985), pp. 33-34.

that it is never invoked. A guarantee that is essentially risk free to the company will be of little or no value to your customers–and may be a joke to your employees.

Why a Service Guarantee Works

A guarantee is a powerful tool–both for marketing service quality and for achieving it–for five reasons.

First, it pushes the entire company to focus on customers' definition of good service–not on executives' assumptions. Second, it sets clear performance standards, which boost employee performance and morale. Third, it generates reliable data (through payouts) when performance is poor. Fourth, it forces an organization to examine its entire service-delivery system for possible failure points. Last, it builds customer loyalty, sales, and market share.

A guarantee forces you to focus on customers. Knowing what customers want is the sine qua non in offering a service guarantee. A company has to identify its target customers' expectations about the elements of the service and the importance they attach to each. Lacking this knowledge of customer needs, a company that wants to guarantee its service may very well guarantee the wrong things.

British Airways conducted a market study and found that its passengers judge its customer services on four dimensions:[1]

1. Care and concern (employees' friendliness, courtesy, and warmth).
2. Initiative (employees' ability and willingness to jockey the system on the customer's behalf).
3. Problem solving (figuring out solutions to customer problems, whether unusual or routine–like multiflight airline tickets).
4. Recovery (going the extra yard, when things go wrong, to handle a particular problem–which includes the simple but often overlooked step of delivering an apology).

British Airways managers confessed that they hadn't even thought about the second and fourth categories. Worse, they realized that if *they* hadn't understood these important dimensions of customer service, how much thought could their employees be giving to them?

A guarantee sets clear standards. A specific, unambiguous service guarantee sets standards for your organization. It tells employees what the company stands for. BBBK stands for pest elimination, not pest control; Federal Express stands for "absolutely, positively by 10:30 A.M.," not "sometime tomorrow, probably." And it forces the company to define each employee's role and responsibilities in delivering the service. Salespeople, for example, know precisely what their companies can deliver and can represent that accurately–the opposite of the common situation in which salespeople promise the moon and customers get only dirt.

This clarity and sense of identity have the added advantage of creating employee team spirit and pride. Mitchell Fromstein, president and CEO of Manpower, says, "At one point, we wondered what the marketing impact would be if we dropped our guarantee. We figured that our accounts were well aware of the guarantee and that it might not have much marketing power anymore. Our employees' reaction was fierce–and it had a lot less to do with marketing than with the pride they take in their work. They said, 'The guarantee is proof that we're a great company. We're willing to tell our customers that if they don't like our service for any reason, it's our fault, not

A service guarantee is valued when a customer's ego is on the line.

theirs, and we'll make it right.' I realized then that the guarantee is far more than a simple piece of paper that puts customers at ease. It really sets the tone, externally and, perhaps more important, internally, for our commitment to our customers and workers."

A payout that creates financial pain when errors occur is also a powerful statement, to employees and customers alike, that management demands customer satisfaction. A significant payout ensures that both middle and upper management will take the service guarantee seriously; it provides a strong incentive to take every step necessary to deliver. A manager who must bear the full cost of mistakes has ample incentive to figure out how to prevent them from happening.

A guarantee generates feedback. A guarantee creates the goal; it defines what you must do to satisfy your customers. Next, you need to know when you go wrong. A guarantee forces you to create a system for discovering errors—which the Japanese call "golden nuggets" because they're opportunities to learn.

Arguably the greatest ailment afflicting service companies is a lack of decent systems for generating and acting on customer data. Dissatisfied service customers have little incentive to complain on their own, far less so than unhappy product owners do. Many elements of a service are intangible, so consumers who receive poor service are often left with no evidence to support their complaints. (The customer believes the waiter was rude; perhaps the waiter will deny it.) Second, without the equivalent of a product warranty, customers don't know their rights. (Is 15 minutes too long to wait for a restaurant meal? 30 minutes?) Third, there is often no one to complain to—at least no one who looks capable of solving the problem. Often, complaining directly to the person who is rendering poor service will only make things worse.

> Without a guarantee, customers won't complain. Or come back.

Customer comment cards have traditionally been the most common method of gathering customer feedback on a company's operations, but they, too, are inadequate for collecting valid, reliable error data. In the first place, they are an impersonal form of communication and are usually short (to maximize the response rate). Why bother, people think, to cram the details of a bad experience onto a printed survey form with a handful of "excellent—good—fair" check-off boxes? Few aggrieved customers believe that completing a comment card will resolve their problems. Therefore, only a few customers—usually the most satisfied and dissatisfied—provide feedback through such forms, and fewer still provide meaningful feedback. As a broad gauge of customer sentiment, cards and surveys are useful, but for specific information about customer problems and operational weaknesses, they simply don't fill the bill.

Service companies thus have a hard time collecting error data. Less information on mistakes means fewer opportunities to improve, ultimately resulting in more service errors and more customer dissatisfaction—a cycle that management is often unaware of. A guarantee attacks this malady by giving consumers an incentive and a vehicle for bringing their grievances to management's attention.

Manpower uses its guarantee to glean error data in addition to allaying customer worries about using an unknown quantity (the temporary worker). Every customer who employs a Manpower temporary worker is called the first day of a one-day assignment or the second day of a longer assignment to check on the worker's performance. A dissatisfied customer doesn't pay—period. (Manpower pays the worker, however; it assumes complete responsibility for the quality of its service.) The company uses its error data to improve both its work force and its proprietary skills-testing software and skills data base—major elements in its ability to match worker skills to customer requirements. The information Manpower obtains before and after hiring enables it to offer its guarantee with confidence.

A guarantee forces you to understand why you fail. In developing a guarantee, managers must ask questions like these: What failure points exist in the system? If failure points can be identified, can their origins be traced—and overcome? A company that wants to promise timely service delivery, for example, must first understand its operation's capability and the factors limiting that capability. Many service executives, lacking understanding of such basic issues as system throughput time, capacity, and process flow, tend to blame workers, customers, or anything *but* the service-delivery process.

Even if workers *are* a problem, managers can do several things to "fix" the organization so that it can support a guarantee—such as design better recruiting, hiring, and training processes. The pest-control industry has historically suffered from unmotivated personnel and high turnover. Al Burger overcame the status quo by offering higher than average pay (attracting a higher caliber of job candidate), using a vigorous screening program (making those hired feel like members of a select group), training all workers for six months, and keeping them motivated by giving them a great deal of autonomy and lots of recognition.

Some managers may be unwilling to pay for an internal service-delivery capability that is above the industry average. Fine. They will never have better than average organizations, either, and they will therefore never be able to develop the kind of competitive advantage that flows from a good service guarantee.

A guarantee uncovers errors—and opportunities to learn.

A guarantee builds marketing muscle. Perhaps the most obvious reason for offering a strong service guarantee is its ability to boost marketing: it encourages consumers to buy a service by reducing the risk of the purchase decision, and it generates more sales to existing customers by enhancing loyalty. In the last ten years, Manpower's revenues have mushroomed from $400 million to $4 billion. That's marketing impact.

Keeping most of your customers and getting positive word of mouth, though desirable in any business, are particularly important for service companies. The net present value of sales forgone from lost customers—in other words, the cost of customer dissatisfaction—is enormous. In this respect, it's fair to say that many service companies' biggest competitors are themselves. They frequently spend huge amounts of money to attract new customers without ever figuring out how to provide the consistent service they promise to their existing customers. If customers aren't satisfied, the marketing money has been poured down the drain and may even engender further ill will. (See the insert, "Maximizing Marketing Impact.")

A guarantee will only work, of course, if you start with commitment to the customer. If your aim is to minimize the guarantee's impact on your organization but to maximize its marketing punch, you won't succeed. In the long run, you will nullify the guarantee's potential impact on customers, and your marketing dollars will go down the drain.

Phil Bressler, owner of 18 Domino's Pizza franchises in the Baltimore, Maryland area, demonstrates the right commitment to customers. He got upset the time his company recorded its highest monthly earnings ever because, he correctly figured, the profits had come from money that should have been paid out on the Domino's guarantee of "delivery within 30 minutes or $3 off." Bressler's unit managers, who have bottom-line responsibility, had pumped up their short-term profits by failing to honor the guarantee consistently. Bressler is convinced that money spent on guarantees is an investment in customer satisfaction and loyalty. He also recognizes that the guarantee is the best way to identify weak operations, and that guarantees not acted on are data not collected.

Compare Bressler's attitude with that of an owner of several nationally franchised motels. *His* guarantee promises that the company will do "everything possible" to remedy a customer's problem; if the problem cannot be resolved, the customer stays for free. He brags that he's paid, on average, refunds for only two room guarantees per motel per year—a minuscule percentage of room sales. "If my managers are doing their jobs, I don't have to pay out for the guarantee," he says. "If I do have to pay out, my managers are not doing their jobs, and I get rid of them."

Clearly, more than two guests of *any* hotel are likely to be dissatisfied over the course of a year. By seeking to limit payouts rather than hear complaints, this owner is undoubtedly blowing countless opportunities to create loyal customers out of disgruntled ones. He is also losing rich information about which of his motels need improvement and why, information that can most easily be obtained from customer complaints. You have to wonder why he offers a guarantee at all, since he completely misses the point.

Why You May Need a Guarantee Even If You Don't Think So

Of course, guarantees may not be effective or practicable for all service firms. Four Seasons Hotels, for example, could probably not get much marketing or operational mileage from a guarantee. With its strong internal vision of absolute customer satisfaction, the company has developed an outstanding service-delivery system and a reputation to match. Thus it already has an implicit guarantee. To advertise the obvious would produce little gain and might actually be perceived as incongruent with the company's prestigious image.

A crucial element in Four Seasons's service strategy is instilling in all employees a mission of absolute customer satisfaction and empowering them to do whatever is necessary if customer problems do occur. For example, Four Seasons's Washington hotel was once asked by the State Department to make room for a foreign dignitary. Already booked to capacity, Four Seasons had to tell four other customers

Maximizing Marketing Impact

The odds of gaining powerful marketing impact from a service guarantee are in your favor when one or more of the following conditions exist:

The price of the service is high. A bad shoe shine? No big deal. A botched $1,000 car repair is a different story; a guarantee is more effective here.

The customer's ego is on the line. Who wants to be seen after getting a bad haircut?

The customer's expertise with the service is low. When in doubt about a service, a customer will choose one that's covered by a guarantee over those that are not.

The negative consequences of service failure are high. As consumers' expected aggravation, expense, and time lost due to service failure increase, a guarantee gains power. Your computer went down? A computer-repair service with guaranteed response and repair times would be the most logical company to call.

The industry has a bad image for service quality–like pest-control services, security guards, or home repair. A guard company that guarantees to have its posts filled by qualified people would automatically rank high on a list of prospective vendors.

The company depends on frequent customer repurchases. Can it exist on a never-ending stream of new triers (like small service businesses in large markets), or does it have to deal with a finite market? If the market is finite, how close is market saturation? The smaller the size of the potential market of new triers, the more attention management should pay to increasing the loyalty and repurchase rate of existing customers–objectives that a good service guarantee will serve.

The company's business is affected deeply by word of mouth (both positive and negative). Consultants, stockbrokers, restaurants, and resorts are all good examples of services where there are strong incentives to minimize the extent of customer dissatisfaction–and hence, negative word of mouth.

with reservations that they could not be accommodated. However, the hotel immediately found rooms for them at another first-class hotel, while assuring them they would remain registered at the Four Seasons (so that any messages they received would be taken and sent to the other hotel). When rooms became available, the customers were driven back to the Four Seasons by limousine. Four Seasons also paid for their rooms at the other hotel. It was the equivalent of a full money-back guarantee, and more.

Does this mean that every company that performs at the level of a Four Seasons need not offer a service guarantee? Could Federal Express, for example, drop its "absolutely, positively" assurance with little or no effect? Probably not. Its guarantee is such a part of its image that dropping the guarantee would hurt it.

In general, organizations that meet the following tests probably have little to gain by offering a service guarantee: the company is perceived by the market to be the quality leader in its industry; every employee is inculcated with the "absolute customer satisfaction" philosophy; employees are empowered to take whatever corrective action is necessary to handle complaints; errors are few; and a stated guarantee would be at odds with the company's image.

It is probably unnecessary to point out that few service companies meet these tests.

External Variables. Service guarantees may also be impractical where customer satisfaction is influenced strongly by external forces the service provider can't control. While everybody thinks their businesses are in this fix, most are wrong.

How many variables are truly beyond management's control? Not the work force. Not equipment

> An airline can't guarantee on-time flights–but it *can* promise courtesy.

problems. Not vendor quality. And even businesses subject to "acts of God" (like weather) can control a great deal of their service quality.

BBBK is an example of how one company turned the situation around by analyzing the elements of the service-delivery process. By asking, "What obstacles stand in the way of our guaranteeing pest elimination?" Al Burger discovered that clients' poor cleaning and storage practices were one such obstacle. So the company requires customers to maintain sanitary practices and in some cases even make physical changes to their property (like putting in walls). By changing the process, the company could guarantee the outcome.

There may well be uncontrollable factors that create problems. As I noted earlier, such things as flight controllers, airport capacity, and weather limit the extent to which even the finest airline can consistently deliver on-time service. But how employees respond to such externally imposed problems strongly influences customer satisfaction, as British Airways executives learned from their market survey. When things go wrong, will employees go the extra yard to handle the problem? Why couldn't an airline that has refined its problem-handling skills to a science ensure absolute customer satisfaction–uncontrollable variables be damned? How many customers would invoke a guarantee if they understood that the reasons for a problem were completely out of the airline's control–if they were treated with warmth, compassion, and a sense of humor, and if the airline's staff communicated with them honestly?

Cheating. Fear of customer cheating is another big hurdle for most service managers considering offering guarantees. When asked why Lufthansa's guarantee required customers to present written proof of purchase, a manager at the airline's U.S. headquarters told me, "If we didn't ask for written proof, our customers would cheat us blind."

A guarantee can generate breakthrough service and change an industry.

But experience teaches a different lesson. Sure, there will be cheats–the handful of customers who take advantage of a guarantee to get something for nothing. What they cost the company amounts to very little compared to the benefits derived from a strong guarantee. Says Michael Leven, a hotel industry executive, "Too often management spends its time worrying about the 1% of people who might cheat the company instead of the 99% who don't."

Phil Bressler of Domino's argues that customers cheat only when *they* feel cheated: "If we charge $8 for a pizza, our customers expect $8 worth of product and service. If we started giving them $7.50 worth of product and service, then they'd start looking for ways to get back that extra 50 cents. Companies create the incentive to cheat, in almost all cases, by cutting costs and not providing value."

Where the potential for false claims is high, a no-questions-asked guarantee may appear to be foolhardy. When Domino's first offered its "delivery within 30 minutes or the pizza is free" guarantee, some college students telephoned orders from hard-to-find locations. The result was free pizza for the students and lost revenue for Domino's. In this environment, the guarantee was problematic because some students perceived it as a game against Domino's. But Bressler takes the view that the revenue thus lost was an investment in the future. "They'll be Domino's customers for life, those kids," he says.

High Costs. Managers are likely to worry about the costs of a service-guarantee program, but for the wrong reasons. Quality "guru" Philip Crosby coined the phrase "quality is free" (in his 1979 book, *Quality Is Free*) to indicate *not* that quality-improvement efforts cost nothing but that the benefits of quality improvement–fewer errors, higher productivity, more repeat business–outweigh the costs over the long term.

Clearly, a company whose operations are slipshod (or out of control) should not consider offering an unconditional guarantee; the outcome would be either bankruptcy from staggering payouts or an employee revolt stemming from demands to meet standards that are beyond the organization's capability. If your company is like most, however, it's not in that shape; you will probably only need to buttress operations somewhat. To be sure, an investment of financial and human resources to shore up weak points in the delivery system will likely cause a quick, sharp rise in expenditures.

How sharp an increase depends on several factors: your company's weaknesses (how far does it have to go to become good?), the nature of the industry, and the strength of your competition, for example. A small restaurant might simply spend more on employee recruiting and training, and perhaps on sponsoring quality circles; a large utility company might need to restructure its entire organization to overcome years of bad habits if it is to deliver on a guarantee.

Even though a guarantee carries costs, bear in mind that, as Crosby asserts, a badly performed service also incurs costs–failure costs, which come in many forms, including lost business from disgruntled consumers. In a guarantee program, you shift from spending to mop up failures to spending on preventing failures. And many of those costs are incurred in most organizations anyway (like outlays for staff time spent in planning meetings). It's just that they're spent more productively.

Breakthrough Service

One great potential of a service guarantee is its ability to change an industry's rules of the game by changing the service-delivery process as competitors conceive it.

BBBK and Federal Express both redefined the meaning of service in their industries, performing at levels that other companies have so far been unable to match. (According to the owner of a competing pest-control company, BBBK "is number one. There is no number two.") By offering breakthrough service, these companies altered the basis of competition in their businesses and put their competitors at a severe disadvantage.

What are the possibilities for replicating their success in other service businesses? Skeptics might claim that BBBK's and Federal Express's success is not widely applicable because they target price-insensitive customers willing to pay for superior service – in short, that these companies are pursuing differentiation strategies.

It is true that BBBK's complex preparation, cleaning, and checkup procedures are much more time consuming than those of typical pest-control operators, that the company spends more on pesticides than competitors do, and that its employees are well compensated. And many restaurants and hotels are willing to pay BBBK's higher prices because to them it's ultimately cheaper: the cost of "errors" (guests' spotting roaches or ants) is higher than the cost of error prevention.

But, because of the "quality is free" dictum, breakthrough service does not mean you must become the high-cost producer. Manpower's procedures are not radically more expensive than its competitors'; they're simply better. The company's skills-testing methods and customer-needs diagnoses surely cost less in the long run than a sloppy system. A company that inadequately screens and trains temporary-worker recruits, establishes no detailed customer specifications, and fails to check worker performance loses customers.

Manpower spends heavily on ways to reduce errors further, seeing this spending as an investment that will (a) protect its market position; (b) reduce time-consuming service errors; and (c) reinforce the company's values to employees. Here is the "absolute customer satisfaction" philosophy at work, and whatever cost increase Manpower incurs it makes up in sales volume.

Organizations that figure out how to offer – and deliver – guaranteed, breakthrough service will have tapped into a powerful source of competitive advantage. Doing so is no mean feat, of course, which is precisely why the opportunity to build a competitive advantage exists. Though the task is difficult, it is clearly not impossible, and the service guarantee can play a fundamental role in the process.

Author's note: I thank Dan Maher for assistance in researching and writing this article.

Reprint 88405

"Actually, I don't want to make a deposit or a withdrawal I just wanted to make sure everything was, you know, fine."

GROWING CONCERNS

The key to growth and profits is getting employees to give away more of your products and services.

My Employees Are My Service Guarantee

by Timothy W. Firnstahl

I own a chain of four restaurants in and around Seattle, and my company exists for one reason only – to make other people happy. Every time a customer leaves one of our restaurants with a more optimistic view of the world, we've done our job. Every time we fail to raise a customer's spirits with good food, gratifying service, and a soothing atmosphere, we haven't done our job.

To the extent that we satisfy customers, we fulfill our company goal. This observation may seem self-evident and trivial – a useful motto, a business axiom that a lot of business-people understandably overlook in the day-to-day flood of details – but I have found it the very key to growth and profits. And after much trial and error, I have come up with a strategy for ensuring customer satisfaction that has worked wonders in our business and can, I'm convinced, work wonders in other businesses as well.

It starts with a guarantee – not that moth-eaten old promise of a cheerful refund – but a guarantee that customers will be satisfied with their whole experience of the company's products and services. It moves on to a system for giving employees complete responsibility and *authority* for making the guarantee stick. It ends with a process for identifying system failures – the problems in organization, training, and other internal programs that cause customer dissatisfaction.

I call the whole thing "ultimate strategy." That may sound pretentious. But because it redefines a company's ultimate reason for being and succeeding, and because it underlines the importance of finding the ultimate causes of every system failure, I think the name is justified.

Service with a smile and a seed of doubt

Ultimate strategy had its origins in the success of a restaurant business I cofounded ten years ago. (I recently started a different restaurant business, but the strategy hasn't changed.) The first restaurant, specializing in steaks and featuring a huge bar, went over so well that we opened another. Five years ago, we had three restaurants, $7.5 million in sales, and moderate profits. Clearly, many of our customers were satisfied.

But I was bothered by what I saw as an unacceptable level of complaints and by our haphazard responses to them. Not that we didn't try. We happily apologized and gave a free dessert to any customer who complained about slow service, and we cheerfully picked up the cleaning bill when one of our employees spilled the soup. Customers who wrote in to complain about reservations mix-ups or rude service got certificates for complimentary meals.

It was just that our procedures for responding seemed all wrong. Giving out that free dessert required approval from a manager. Getting a suit cleaned meant filling out a form and getting a manager to sign it. I also didn't like the idea that people had to write us with their complaints before we made amends. And I wasn't convinced that a free meal was enough.

Moreover, our response to complaints didn't appear to have any effect on the number or type of complaints we received, most of which concerned speed of service and quality of food. And it wasn't the employees' fault. They knew complaints had top priority, but they didn't know how to respond to them. We were all on a treadmill, getting nowhere.

The guarantee

Then five years ago, when the book *In Search of Excellence* was in vogue, I spent considerable time writing Ten Tenets of Excellence for our organization. We included them in our train-

Founder and CEO of Satisfaction Guaranteed Eateries, Inc. in Seattle, Timothy W. Firnstahl is a restaurant zealot and the fourth generation of his family in the food industry. Previously, he wrote for HBR on how entrepreneurs can delegate responsibility.

ing manuals and posted them in the restaurants and the offices. One day about a year later, someone asked me what the sixth tenet was, and I couldn't tell her. It came to me that if I couldn't remember the Ten Tenets of Excellence, surely no one else could either. That meant the company had no strategy known to its employees.

So I hit on something simpler and more compelling–the guarantee. We expressed it as a promise: *Your Enjoyment Guaranteed. Always.* As a company rallying cry, it seemed to work much better than the Tenets of Excellence. Cryptic mission statements, unreviewed strategic plans, the hidden dreams of management: all that gave way to a company game plan–customer satisfaction–that everyone could understand and remember and act on. For the first time, employees and management had a strategy in common.

Your Enjoyment Guaranteed. Always. This promise became our driving force. We included it in all our advertising. We printed it on every menu, letterhead, and guest check. To make it live for our employees, we did a series of internal promotions. We reduced it to an acronym, YEGA, and posted it everywhere for employees to see.

> What good is a guarantee that makes complaining an ordeal for the customer?

We held a series of meetings, where we found workers receptive to both the acronym and the simplicity of the idea. Each of our 600 employees signed a contract pledging YEGA follow-through. We created a YEGA logo and put it everywhere, on report forms, on training manuals, on wall signs. We started the *YEGA News* and distributed YEGA pins, shirts, name tags, even underwear. We announced that failure to enforce YEGA would be cause for dismissal.

For a year or so, YEGA dominated the company's consciousness. But as time went by, I grew increasingly uncomfortable. Complaints were coming in at the same old rate. I could see the guarantee being implemented here and there, now and then, but not on a regular, companywide basis. I'd run into another brick wall.

Empowering employees

One evening about two years back as I was driving home from work, the cause of the problem hit me. The guarantee by itself wasn't enough. We had given employees responsibility without giving them authority. The result was that they tried to bury mistakes or blame others. I saw it every time we tried to track down a complaint. The food servers blamed the kitchen for late meals. The kitchen blamed the food servers for placing orders incorrectly.

Problems inevitably crop up in a busy restaurant, and when a customer grumbles the tendency is to gloss over the complaint with pleasantness. Follow-through means fetching the manager or filling out forms or both. Climbing the ladder of hierarchical approvals is simply too frustrating and time-consuming –for customer as well as employee.

For our guarantee to be truly effective, we needed to give workers themselves the power to make good on the promise of the guarantee–at once and on the spot. Eliminate the hassle for the customer and for ourselves. No forms to fill out, no phone calls to make, no 40 questions to answer, just immediate redress by the closest employee.

So I instituted the idea that employees could and should do *anything* to keep the customer happy. In the event of an error or delay, any employee right down to the busboy could provide complimentary wine or desserts, or pick up an entire tab if necessary.

Of course, we provided some guidelines. For instance, when guests have to wait more than 10 minutes beyond their reservation time, but less than 20, we suggest free drinks. If they wait more than 20 minutes, the entire meal might be free. If the bread arrives more than 5 minutes after the guests sit down, we suggest free clam chowder. And so forth, using what we know to be optimum intervals for most orders.

At the same time, we urged employees not to get bogged down in the guidelines. The last thing we wanted was nitpicking: "OK, I got them the bread in five minutes exactly. Do I just apologize, or do they get clam chowder?" Satisfaction does not mean quibbling–it means a contented customer. Different guests respond in different ways, so we told our employees not to feel limited by the guidelines and to do whatever it took to make sure guests enjoyed themselves.

Employees were initially wary of their new authority. Never having had complete control, they were naturally hesitant and skeptical. It was hard to convince them they wouldn't be penalized for giving away free food and drinks.

But once they got used to the idea, employees liked knowing that the company believed so strongly in its products and services that it wholeheartedly stood behind its work–and theirs. They liked working for a restaurant known for its unhesitating commitment to customer satisfaction. Preeminence in any field gives people feelings of self-worth they could never get from just making a buck. Their power as company representatives increased their pride in the business, and that, in turn, increased motivation.

Once our employees overcame their skepticism, they quickly grew creative and aggressive in their approach to the guarantee. In one case, a customer wanted a margarita made the way a competitor made its. So our bartender called the bartender at the other restaurant and, bartender-to-bartender, learned the special recipe. In another case, an elderly woman who had not been in our restaurant for years ordered breakfast, which we no longer served. The waiter and the chef sent someone to the market for bacon and eggs and served the breakfast she wanted.

If the guarantee is really working the way it's supposed to, customers become less inhibited about complaining. Too often, customers hold their peace but vote with their feet by taking their business to the competition. The promise of the guarantee's enforcement stimulates them to help us expose our own failures.

We even asked for their criticism. Once a month, using reservations lists and credit card charges, groups of employees called several hundred customers and asked them to rate their experience. Were the food and service lousy, OK, good, very good, or excellent? If they said "OK," that meant "lousy" to us, and they got a letter of apology, a certificate for a free meal, and a follow-up phone call.

Aside from the data we gathered, the phone calls were great promotion. Most people were amazed and delighted that we took the trouble to phone them, and many developed enormous loyalty to our restaurants.

System-failure costs

Customer complaints are company failures and require immediate correction. So far so good. But corrections cost money. Free drinks and meals add up quickly.

Yet, paradoxically, spending money is the goal. Every dollar paid out to offset customer dissatisfaction is a signal that the company must change in some decisive way. The guarantee brings out a true, hard-dollars picture of company failures and forces us to assume full responsibility for our output. The cost of keeping a company's promises is not just the price tag on the guarantee, it is the cost of system failure. The money was spent because the product did not perform, and when the product fails to perform, the system that produced it is at fault.

A somnolent business can be rudely awakened by the magnitude of its system-failure costs. We certainly were. Our previous guarantee expenses doubled. The problems had always been there, hidden. Only the huge cost of the new strategy revealed that they were gutting profits. Suddenly, we had a real incentive to fix the systems that weren't working, since the alternatives – sacrificing profits permanently or restricting the power to enforce the guarantee – were both unacceptable.

> Every dollar you give away is a plus – it puts your finger on a problem you can fix.

Notice that system-failure costs are not the same as employee-failure costs. System-failure costs measure the extent of the confusion in company structure, for which management alone is to blame. By welcoming every guarantee payoff – every system-failure expense – as an otherwise lost insight, you can make every problem pay a dividend. The trick is to reject Band-Aid solutions, to insist on finding the ultimate cause of each problem, and then to demand and expect decisive change. (Another way to sugarcoat the pill of system-failure costs is to think of the free food and drinks as a word-of-mouth advertising budget. No one forgets to mention a free meal to a friend or neighbor.)

Our search for the culprit in a string of complaints about slow food service in one restaurant led first to the kitchen and then to one cook. But pushing the search one step further revealed several unrealistically complex dishes that no one could have prepared swiftly.

In another case, our kitchens were turning out wrong orders at a rate

The Hassle Factor

Imagine you've bought a new pair of shoes at a downtown store. A week later, one sole starts to come off, so you take them back. You drive downtown through heavy traffic and spend 15 minutes finding a parking place. You explain the problem to the salesman, who says, "We stand behind our merchandise." He gives you a new pair of shoes.

Question One: Are you happy?

Answer: Well, no, you're not. Sure, you got a new pair of shoes, and the salesman was pleasant enough, but you had to take time out of your day and go to a lot of trouble to get what you should have gotten in the first place. In short, the whole transaction was a hassle, and neither the salesman nor the store did anything to make it up to you.

Question Two: What should they have done?

Answer: Replace plus one. Besides giving you a new pair of shoes, the salesman should have thrown in a pair of socks or stockings to repay you for your hassle. Instead of, "We'll replace inferior merchandise whenever a customer complains," the store's message should be, "We really regret your inconvenience and want to make you happy."

Like the shoe store, we stand behind our products and services. Unlike the shoe store, we'll do more than the customer demands to make it right. If a guest doesn't like her salad, don't charge her for it. But what about the Hassle Factor?

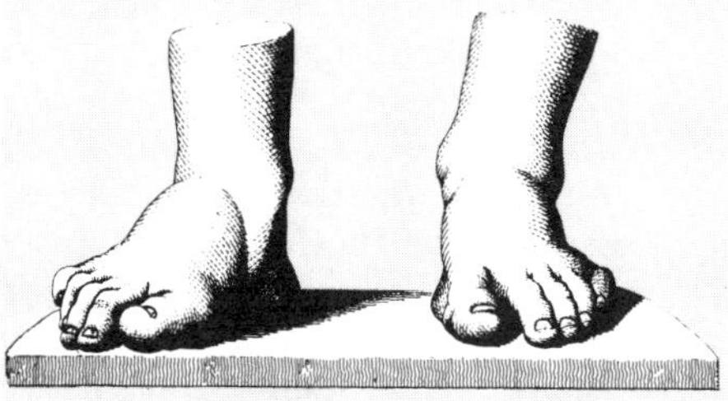

Replace plus one. By all means, give her the salad free of charge. But buy her a drink or dessert as well – or whatever else it takes to make her happy.

Adapted from the restaurant training manual

that was costing us thousands of dollars a month in wasted food. The cooks insisted that the food servers were punching incorrect orders into the kitchen printout computer. In times past, we might have ended our search right there, accused the food servers of sloppiness, and asked everyone to be more careful. But now, adhering to the principle of system failure not people failure, we looked beyond the symptoms and found a flaw in our training. We had simply never taught food servers to double-check their orders on the computer screen, and the system offered no reward for doing so. Mistakes plummeted as soon as we started training people properly and handing out awards each month for the fewest ordering errors and posting lists of the worst offenders (no punishments, just names).

Of course, correcting system failures is seldom an easy task. One way to avoid making problems worse is to audition problem solutions with small, quick-hit field tests. For example, we experimented with new service procedures at one station in one restaurant, or we offered new menu items as nonmenu specials, or we borrowed equipment for a test run before leasing or buying it. When we had a problem with coffee quality, we tried using expensive, high-quality vacuum carafes in one restaurant. Quality improved substantially (and waste was cut in half), so we adopted the thermoses in all our restaurants.

What is a word-of-mouth advertising budget? It's the money you spend honoring your guarantee.

When some customers complained about our wine service, we realized that we gave the subject only three pages in our employee manual. So we put together a training and motivation package that included instruction about the characteristics and distinctions of different wines, as well as a system of awards for selling them effectively. We also pointed out to our food servers that selling more wine increases the size of checks and thus of tips.

In short, honoring the guarantee has led to new training procedures, recipe and menu changes, restaurant redesign, equipment purchases, and whatever else it took to put things right and keep them right. In the long run, the guarantee works only if it reduces system-failure costs and increases customer satisfaction.[1]

This kind of problem solving is popular with employees. Since the object of change is always the company, employees don't get blamed for problems beyond their control.

As you find and correct the ultimate causes of your system failures, you can reasonably expect your profits to improve. But you can begin to tell if you're succeeding even before you see it on the bottom line. Remember, costs will go up before they come down, so high system-failure costs and low phone-survey complaint rates probably mean you're on the right track. Conversely, low system-failure costs and a high rate of "lousies" and "OKs" from customers almost certainly indicate that the promise is not being kept, that your expensive system failures are not getting corrected, and that your organization has yet to understand that customer satisfaction is the only reason for the company's existence.

Our own system-failure costs rose to a high of $40,000 a month two years ago and then fell to $10,000 a month. Meanwhile, sales rose 25%, profits doubled, and the cash in the bank grew two-and-a-half times.

Making it work

It is easier to give someone a bowl of clam chowder than a free CAT scanner or an industry marketing study, so of course the nature of the guarantee will change from business to business. Still, the point is not free food, the point is customer satisfaction. It is always possible to satisfy the customer if the business is sufficiently committed to that goal.

Here are my suggestions for formulating your own ultimate strategy.

1. *Make the guarantee simple and easy to understand.* Think about the company's primary customer benefit and how you achieve it. In our case, the principal benefit is enjoyment. For many, it will be dependability. For others, cost or flexibility.

Firnstahl's first rule: Always deal with complaints before they're made.

For the sake of impact, try to develop a guarantee that's memorable, maybe one that reduces to an acronym. The restaurants I now own use WAGS (We Always Guarantee Satisfaction), which I like even better than YEGA. Whatever you do, make it significant, simple, and unconditional. Think of these famous promises that changed whole companies:

"We try harder." (Avis)

"Absolutely Positively Overnight." (Federal Express)

Once you settle on a guarantee, commit to it for the long term. Continual change confuses the public and the organization. Plan to stick with a particular promise for at least five years.

2. *Make sure employees know how to use their new authority.* For most employees, full power and responsibility to put things right will be a new experience. After all, they're used to the old hierarchical approach. So it's up to you to make sure they don't underuse their power. In our training programs, we advise new employees to take action before the guest has to ask for a remedy. We don't want to make customers decide whether they're entitled to get something free – most people find that embarrassing. The food server should find the solution and present it to the guest as a done deal: "I'm sorry your drink wasn't prepared the

1. For more on guarantees, see Christopher W. L. Hart, "The Power of Unconditional Service Guarantees," HBR July-August 1988, p. 54.

way you like it. Of course, there will be no charge for that. And please accept these chowders on the house with my apologies."

We also insist that the customer is always right, even when the customer is wrong. Let's say a guest insists that all clam chowder has potatoes. He's wrong, but that's no excuse to make him look stupid. When we say, "The guest is never wrong," we mean a server should never question a guest's judgment and perception. Don't stand and argue about whether a steak is medium-rare or medium. Take it away and get one broiled the way the customer wants it.

> Giving your employees total power and authority will scare them to death and make you rich.

The real issues are these: The guest is there to have a good time. The guest is in the employee's care.

Finally, we think that power and responsibility are not enough. Employees must also have rewards. Good thinking and positive action deserve money, praise, the limelight, advancement, and all the other encouragements a company can think of.

We spark employee thought and action by dividing a $10,000 bonus among the employees of each restaurant once its system-failure costs and phone-audit complaint rates drop to 25% of their all-time highs. Every month, we pay thousands of dollars in awards to employees who have helped to find and cure the ultimate causes of system failures. In effect, we commission everyone to change the organization for the better.

3. *Make progress visible.* Stay away from written progress reports – graphs communicate better. A creative in-house accountant can play with the data until it's readily understandable to everyone. We display our new WAGS graphics throughout the company for everyone to see.

In our experience, system-failure costs go through four phases.

Start: Employees are wary of using their new power and authority. Phone-audit complaint rates are high and system-failure costs are low.

Under way: Employees begin to believe in the organization's commitment to the guarantee. Phone-audit complaint rates are still high; system-failure costs start to rise.

Mid-point: Employees accept and act on the company promise. System-failure costs remain high. Phone-audit complaint rates start dropping as the company starts satisfying customers in earnest.

Success: The company has achieved elemental change and raised itself to a higher level of merit. System-failure costs and phone-audit complaint rates are both low.

In general, there is a roller coaster effect that tells you when ultimate strategy is working. Costs go up. Complaints go down. Sales go up. Costs go down. Profits go up.

One word of caution: you will never perfect your company's system. As long as you offer an absolute guarantee on your products and services, you will incur system-failure costs. There is always more work to do, and a CEO's personal commitment and persistence are often necessary to get it done. But motivated employees are essential.

People often ask us where we find such wonderful employees. While it's true that we screen carefully, I believe our employees are better than most because they have the power and the obligation to solve customer problems on their own and on the spot. Giving them complete discretion about how they do it has also given them pride. Many companies have tried so many different programs and gimmicks that employees have become cynical and indifferent. The people who work for us know we take our guarantee seriously – and expect them to do the same. We use the same ultimate strategy to satisfy both customers and employees.

Reprint 89407

Adding Value to Products

SPECIAL REPORT

Focus on what gives your company its competitive edge. Outsource the rest.

Beyond Products: Services-Based Strategy

by James Brian Quinn, Thomas L. Doorley, and Penny C. Paquette

When communication was limited to telephones and letters, and transportation took weeks or months instead of hours or days, concentrating on a few products – and the vertical integration that let managers control every step of their production processes – made real sense. Now such traditional strategic formulas no longer hold.

Thanks to new technologies, executives can divide up their companies' value chains, handle the key strategic elements internally, outsource others advantageously anywhere in the world with minimal transaction costs, and yet coordinate all essential activities more effectively to meet customers' needs. Under these circumstances, moving to a less integrated but more focused organization is not just feasible but imperative for competitive success.

Companies that understand this new approach – Honda, Apple, and Merck among them – build their strategies not around products but around deep knowledge of a few highly developed core service skills. In such companies, the organization is kept as lean as possible. The company strips itself down to the essentials necessary to deliver to customers the greatest possible value from its core skills – and outsources as much of the rest as possible. As a result, management focuses on what it does best, avoids distractions, and leverages its organizational and financial resources far beyond what traditional strategies allow.

The power of services

To rethink their strategies objectively, managements need to break out of the mind-set that considers manufacturing (or goods production) as separate from (and somehow superior to) the service activities that make such production possible and effective. In fact, most companies – product manufacturers and service providers alike – are largely service operations. U.S. employment and output statistics reflect this clearly, as do internal company cost figures. Upon analysis, so do most manufacturers' value chains.

First consider some numbers. Approximately 76% of all U.S. workers are employed in industries commonly thought of as services: communications, transportation, health care, wholesale and retail distribution, financial services, and professional firms. Of those working in manufacturing industries, 65% to 75% perform service tasks ranging from critical production-related activities like research, logistics, maintenance, and product and process design to indirect staff services like accounting, law, financing, and personnel. Overall, services account for over three-fourths of all costs in most U.S. industries.

The role of services in providing value is ever more important. Not long ago, most of a product's value added came from the production processes that converted raw material into useful forms (steel into auto bodies, for example, or grain into edible cereals). Now, however, value added is increasingly likely to come from technological improvements, styling features, product image, and other attributes that only services can create.

This occurs in part because systemization and automation have driven production costs steadily downward, thus diminishing their

James Brian Quinn is the William and Josephine Buchanan Professor of Management at the Tuck School at Dartmouth. This is his ninth contribution to HBR. Thomas L. Doorley is a founder and managing partner of Braxton Associates, an international strategy consulting firm. Penny C. Paquette is research associate and director of the International University of Japan Program Office at the Tuck School.

relative importance in most companies' value chains. More and more companies are beginning to look like those in the personal computer industry, where producing the actual "box" is a low-margin activity, and software and service-support activities create most of a product's

Manufacturers and service providers alike are largely service operations.

value to customers. In fact, nonmanufacturing services are so preeminent in many pharmaceutical, clothing, food, and sports equipment companies (to name just a few) that even classifying them as manufacturers seems open to question.

In pharmaceuticals, for example, value is added primarily by service activities like drug development in R&D, carefully constructed patent and legal defenses, thorough clinical and regulatory clearances, and effective drug detailing and distribution systems. With production costs a trivial portion of a drug's value, leading companies' strategies concentrate on specialized activities within the value chain. Merck, for example, focuses on a powerful research-based patent position, while Glaxo targets rapid regulatory clearance of drugs.

SCI Systems, the world's largest electronics subcontractor, is a less obvious case in point. SCI produces a wide variety of communications, computer, and advanced instrumentation equipment that it sells both to value-added resellers and on an OEM basis. The company has been growing at a rate of 35% per year in its cutthroat business by outsourcing as much of its production processes and nonessential services as possible. Management concentrates its resources on design and development, logistics management, quality control, and the company's special expertise – low-cost assembly technologies for surface mounting components on both sides of the circuit board. The thin overhead structure such outsourcing allows (130 managers for 7,000 employees) lets SCI respond flexibly, rapidly, and precisely, with lower bureaucratic costs than its competitors or its customers themselves would incur.

In addition to enhancing the leverage of services inside companies, new technologies have significantly increased the relative power of service enterprises. For both manufacturers and service companies, this presents important opportunities and challenges.

On the one hand, services technologies offer astute managers new options for lowering costs, restructuring their organizations, and redefining their strategic focus. On the other hand, those who ignore the opportunities that services technologies create will surely sacrifice competitive advantages to their more farsighted rivals. Our three-year study of technology's impacts provided us with examples of both phenomena time and time again. (The insert, "The Impact of Services Technologies," describes this research.)

At the most basic level, technology has made it possible for independent companies to specialize in particular service activities, automate them, and create higher value added at lower costs than all but a few integrated companies can attain. Automatic Data Processing provides a good example.

ADP developed a flexible payroll-handling system that is so low cost it now processes payrolls for more than 10% of its potential client base. From this initial service, ADP expanded into routine bank accounting and tax filings for its customers and their employees, then into ERISA reporting, personnel records, and financial analyses, and finally into personalized communications like printing slogans, messages, or logos on checks and including notes with employees' paychecks. ADP's customers get expertise they could not afford in-house, lower overhead costs thanks to ADP's substantial economies of scale, and an objective check on their own costs for these services.

Similarly, the $1.5 billion ServiceMaster uses its system economies and specialized management skills to raise the quality of its customers' maintenance services while at the same time lowering their costs. ServiceMaster's database, which covers 14 years of maintenance history on 17 million pieces of equipment at thousands of locations, enables the company to determine objectively how its customers' facilities should be maintained, when equipment purchases and preventive maintenance will pay off, and when parts should be replaced. So effective are its systems that ServiceMaster and its customers often invest jointly in new equipment and share the resulting productivity gains.

As these examples suggest, many service companies have become large, capital-intensive, technology-dominated organizations on which companies from almost every industry depend for specialized knowledge and assistance. Service companies are now among the most important suppliers, customers, market innovators, coalition partners, and repositories of market knowledge available to manufacturers.[1] A prime illustration is Toys "R" Us, which sells at year-round discount prices

Merck focuses on a powerful patent position, Glaxo on regulatory clearance of drugs.

the largest selection of toys available nationwide.

To maintain inventories of 18,000 items in each store at lowest cost, Toys "R" Us must make huge off-season purchases, respond rapidly to developing fads, yet minimize systemwide storage and handling costs. It does this through what is perhaps the most detailed materials management and movement system in re-

1. James Brian Quinn, Jordan J. Baruch, and Penny C. Paquette, "Exploiting the Manufacturing-Services Interface," *Sloan Management Review*, Summer 1988, p. 45.

tailing, combining daily updates from its electronic point-of-sale system with an expert system linked to its sophisticated supplier information models.

The power of this system does not stop with Toys "R" Us. The 350-store system is so effective that the company knows more about toy sales than any of its manufacturer-suppliers, who often use their spring and summer Toys "R" Us sales to gauge production levels and to redesign products for their Christmas lines. With its market share approaching 20%, Toys "R" Us is so powerful that it can strongly influence product offerings and require manufacturers to meet its needs (for example, providing special packaging and four-sided product markings for its open, high-stack displays).

Shifting strategic focus

True strategic focus means that a company can concentrate more power in its chosen markets than anyone else can. Once, this meant owning the largest resource base, manufacturing plants, research labs, or distribution channels to support product lines. Now physical facilities – including a seemingly superior product – seldom provide a sustainable competitive edge. They are too easily bypassed, reverse engineered, cloned, or slightly surpassed. Instead, a maintainable advantage usually derives from outstanding depth in selected human skills, logistics capabilities, knowledge bases, or other service strengths that competitors cannot reproduce and that lead to greater demonstrable value for the customer.

Recognizing this, smart strategists no longer analyze market shares and their associated cost positions ad infinitum, nor do they build integrated companies to exploit them. Instead, they concentrate on identifying those few core service activities where their company has – or can develop – unique capabilities. Then they aggressively seek ways to eliminate, limit, or outsource activities where the company cannot attain superiority, unless those activities are essential to its chosen areas of strategic focus.

The Impact of Services Technologies

In a three-year study begun under the auspices of the National Academy of Engineering, we explored the important impacts of technology in the service sector. Following the dictionary, we defined technology broadly as the systematic application of knowledge – particularly knowledge of physical, chemical, and systems phenomena – to useful purposes. We studied not just information technologies but all the other systems, operations, software, and hardware technologies developed specifically for service functions or applied in them. Among those included were: the diagnostic techniques, treatment devices, and specialized procedures developed for health care; advanced cargo handling and passenger movement equipment; specialized food distribution and preparation systems for restaurant chains; and automatic teller machines and satellite communication systems used in banking.

We examined the way such technologies are affecting industry structures, corporate strategies, and organizations. Our findings are based on an analysis of all relevant databases (including the Strategic Planning Institute's PIMS unit, Standard & Poor's COMPUSTAT tapes, as well as databases from the U.S. Departments of Commerce and Labor), a series of case studies, and numerous interviews with key technical and senior managers in companies identified as outstanding developers or users of services technologies. We gratefully acknowledge the support of the following companies in financing this project: Bell & Howell, Bell Atlanticom, Bankers Trust, Royal Bank of Canada, Braxton Associates, and American Express.

Apple Computer illustrates the value of this kind of activity-based focus. Like many other now-large companies, Apple initially succeeded by organizing itself as an intellectual holding company that purposely manufactured as little internally as possible. Because its business strength lay in creating the friendly look and feel of its software and hardware, Apple's management concentrated on designing and controlling its products' concepts, appearance, and key software, especially Apple DOS, which was not even made available for license. Other components and activities were outsourced wherever possible.

Apple bought microprocessors from Synertek, other chips from Hitachi, Texas Instruments, and Motorola, video monitors from Hitachi, power supplies from Astec, and printers from Tokyo Electric and Qume. (The Apple II was estimated to cost less than $500 to build, of which $350 was for purchased components.[2]) Similarly, Apple kept its internal service activities and investments to a minimum by outsourcing application software development to Microsoft, promotion to Regis McKenna, product styling to Frogdesign, and distribution to ITT and ComputerLand.

While this strategy may have been essential in Apple's early days, when it lacked both the time and the capital to build factories or hire a sales force, even today it is structured less like a traditional manufacturing company and more like a $4 billion service company that happens to

2. William H. Davidson and Edward F. Colby, Jr., "Apple Computer, Inc.," University of Virginia Darden Graduate Business School Foundation Case Study, 1983 (UVA-BP219).

have limited manufacturing facilities. Yet it has produced spectacular employee productivity and profitability: in 1988, Apple had $377,000 in sales per employee and $57,400 in profits per employee, outperforming the more vertically integrated IBM ($154,200 and $22,600) and DEC ($101,900 and $10,600).

To develop an activity-focused strategy, management needs to concentrate its competitive analyses not on market share but on the relative potency of the services undergirding its own and its competitors' product positions. Too much attention has been paid to high market share, especially since it often is bought by inappropriate pricing and short-term marketing strategies. High market share and high profitability together come from having a high activity share, or the most effective presence in selected services that the market truly desires.

Defining each activity in the value chain as a service that can be either produced internally or sourced externally is the first step in this new competitive analysis. Next management must ask a series of questions about each activity: Do we have or can we achieve best-in-the-world capabilities for this service? If so, should we make it a part of our core strategy? If not, what possibilities exist for outsourcing the activity or forming a strategic alliance with someone who does have superior capabilities? Finally, management needs to focus the organization's energies on two sets of activities: those where it can create unique value and those it must control to maintain its supremacy in the critical elements of its value chain.

In recent years, outsourcing has been disparaged for "hollowing out" companies and making them uncompetitive. Approached strategically, as we suggest, the effects of outsourcing are just the opposite. Whenever a company produces something internally that others can buy or produce more efficiently and effectively, it sacrifices competitive advantage. Conversely, the key to strategic success for many companies has been a carefully developed coalition with one or more of the world's best suppliers, product designers, advertising agencies, distributors, financial houses, or other service providers. Apple is one company that profits from such an approach. Honda Motor Company is another.

Soichiro Honda's personal commitment and engineering skills led him to create the world's finest team for designing small, efficient engines, first for bicycles, then for motorcycles. In the company's early years, limited capital resources and the cost-competitiveness of the motorcycle market forced Honda to develop very small, efficient assembly operations and to outsource as much of its nonengine fabrication work as possible. At the same time, Honda's president, Takeo Fujisawa, used innovative financing to build the strongest motorcycle distribution network in Japan.

Once Honda's motorcycles had succeeded at home, the company parlayed these same core service capabilities into building a steadily broader line of motorcycles, automobiles, and household machines sold worldwide. Its successes in the United States and elsewhere are still based on its unique depth in small engine design, moderate-scale assembly management, superb outsourcing logistics, and creative marketing and distribution coalitions. It has avoided many of the investments and inflexibilities that its major competitors have been saddled with, while building itself a number one share of the U.S. foreign car market and a number three share of all cars sold in Japan.

While both Apple and Honda began by outsourcing critical activities, other companies have adopted this strategy later in life. Intel Corporation, for example, at first relied heavily on its internal production prowess to lower costs, increase reliability, and open new markets. Now it focuses primarily on its extraordinary competencies in chip design and development. When its chips move toward commodity status, Intel often withdraws from the manufacturing process, subcontracting that to silicon foundries and other producers that concentrate on volume manufacturing and may have inherent cost advantages.

Similarly, in a lower tech industry, E&J Gallo Winery shifted to outsourcing an activity many vintners consider the core of their business–growing the specialized grapes for its wines. Gallo devotes its resources and management attention to maintaining the legendary marketing and sales strengths that give the company its volume advantages and to using the deep knowledge base its 31% market share can provide to purchase grapes with the precise qualities its wines require. An important side benefit is that Gallo has avoided the investments and risks inherent in growing grapes.

What stays, what goes?

The great part of most companies' costs (other than those for purchased materials) typically occur in overhead categories. Even in manufacturing, more than two-thirds of all nonmaterial costs tend to be indirect or overhead expenses. Yet most overhead is merely services that the company began to buy internally. Reconceptualizing internal staffs and overhead costs as services that could be bought externally exposes a powerful, unexploited source of competitive advantage for most companies. Management can ask activity by activity: Are we really competitive with the world's best here? If not, can intelligent outsourcing or coalitions improve our productivity and long-term competitive position?

Competitive analyses should consider not just competitors in the company's own industry or the exclusive providers of a service (like insurance companies or communications companies) but *all* the potential providers and industries that might cross-compete in the activity, using "best in the activity" as the relevant benchmark. This approach broadens the analytical process and gives it a much more external, market-driven orientation. It also introduces new objectivity into the evaluation process and, at a minimum, generates strong pressures for productivity gains if management decides to continue sourcing internally. In many cases, managers discover that specialized sources are so much more cost-effective than their internal groups that they outsource activities long considered integral to their business.

Clearly, however, not all overhead is worthless–and therefore to be eliminated. Many of the most essential activities that modern companies perform are in what could be considered overhead categories. The objective is not to slash costs simplistically but rather to produce or buy needed services most effectively.

By limiting or shedding activities that provide no strategic advantage, a company can increase the value it delivers to both customers and shareholders and, in the process, lower its costs and investments. Conversely, if a company performs activities that it could buy more effectively, costs tend to balloon, and it loses competitive advantage. The company may also become more vulnerable to takeovers by outsiders who see in these activities the potential for margin gains.

Determining which core activities to emphasize is not always easy or obvious. Often a company's true strengths are obscured by management's tendency to think in divisional or product terms and by each functional group's need to see itself as the main source of the enterprise's success. But careful analysis will identify the few critical activities that drive (or could drive) the company's strategy and that it therefore must dominate.

Consider recent developments in the credit card industry. Processing customer charges and transactions is obviously an important activity in this business. Yet both MasterCard and Visa have clearly decided that their distinctive advantage lies not in this costly and complicated activity–which they often outsource to arch-competitor American Express–but in their large customer bases, retail networks, and ability to market effectively to both.

> To control its manufacturing process, Honda does its engine-related R&D in Japan.

Managers must, of course, plan and control outsourcing so that the company does not become overly dependent on–and hence dominated by–its partners. In most cases, this means consciously developing and maintaining alternative competitive sources. It also means the company may have to perform activities at certain critical stages of its value creation process that it would otherwise outsource if efficiency were the sole criterion.

As Apple demonstrates, a company need not fabricate many components itself to control manufacturing. But it must dominate the strategic steps in the manufacturing process and develop technological knowledge, logistics systems, and commercial intelligence networks about the other steps that are second to none. That is why Honda still manufactures key engine parts in Japan and does all of its critical engine-related R&D there, although it outsources body parts overseas and shares responsibility for body design with its affiliates worldwide.

One way to avoid becoming too dependent on critical suppliers is to seek alliances with outstanding but noncompeting enterprises. Seagate Technology, the world's leading manufacturer of hard disk drives for personal computers, provides a good example. To gain competitive advantage against Japanese rivals, Seagate decided to absorb the cost of delivering its products to customers and to guarantee delivery within four days. To handle this crucial service, the company developed a special relationship with Skyway Freight Systems to make deliveries from its California distribution center to customers all over the United States. On a per unit basis, Seagate's payments to Skyway amounted to less than 1% of each disk drive's total cost. Yet in six months and thousands of shipments, only three deliveries arguably were late.

On rare occasions, companies can even outsource some strategic components to potential competitors – as Apple did with its microprocessor chips – provided that they develop and protect other key elements in the value creation process and control the critical information linkages between their various suppliers and customers – as Apple did by controlling Apple DOS and its marketing interfaces.

This kind of refocusing and strategic hollowing out goes well beyond mere tactical redeployment for efficiency's sake. By avoiding investments in vertical integration and by managing "intellectual systems" instead of workers and machines, companies decrease total investment and leverage resources substantially. They also minimize certain unavoidable risks.

For example, if one unit in an outsourced system underperforms, management can substitute another competitor's components or services more quickly than it could shut down an internal operation or develop an alternative itself. Similarly, if new technologies suddenly appear, it is easier to switch among outside sources than to forsake internal investments. Lastly, if there is a cyclical or temporary drop in demand, the coordinating company is not saddled with all the idle capacity and inventory losses.

With strategic outsourcing, companies have at their disposal the world's best talent, offering both higher quality and greater flexibility than internal groups could provide. For this reason, many large manufacturers now rely on smaller, state-of-the-art product and service suppliers to shorten product-development cycle times. While the larger company must typically maintain strong R&D efforts in its core technologies – and be willing to share its longer term plans with suppliers to ensure that its new products make the best use of suppliers' designs – it can substantially leverage its R&D funds through inputs from specialized technical vendors. Dramatic savings, totaling some 70% to 80% of the cost of key components or designs, are common when joint development is managed properly.

In the early 1980s, General Motors, like many other large companies, realized that the rigidities and costs of its internal suppliers were not only raising its cost per unit against its competitors' but also cutting off the company from the R&D support and innovations that outside suppliers could provide. In a policy reversal, GM began to encourage more extensive outsourcing and, among other initiatives, turned to PPG Industries to handle the painting of auto bodies. Today PPG manages and operates the paint shops in GM assembly plants, lowering GM's cost per unit and offering the latest advances in robotics and materials research related to painting.

Today PPG manages and operates the paint shops in GM assembly plants.

Perhaps most important, strategic outsourcing allows a company to focus all its attention on the areas in which it adds the most value per unit of input. This creates entry barriers for would-be competitors in a variety of ways. Because they are leaner, these companies can respond faster to customers and move more easily to form the cross-functional teams that best provide innovative products for new market needs. Because they have fewer management levels, less bureaucracy, and more opportunities for employees' personal and entrepreneurial development, they can often attract better people. Finally, they can leverage managerial talent – their most limited and most crucial resource – because their executives no longer need to waste time and attention on peripheral activities.

Service-driven competition

Global sourcing, volatile currency fluctuations, fast-changing communications and information technologies, and the increased leverage of services are combining to restructure entire industries as well as individual companies. As an example, consider biotechnology.

The industry is becoming a loosely structured network of service enterprises built around specialized core competencies and joined together (often temporarily) for one undertaking, while remaining suppliers, competitors, and customers in others. Many research groups and companies focus only on identifying and patenting biological organisms (often proteins) at the laboratory level. Others develop and license cell lines that can reproduce these organisms. Still others have developed processes for using these cell lines to make particular proteins in sufficient quantities for clinical tests and commercialization. Some enterprises specialize in running clinical trials, others concentrate on having the marketing expertise and distribution channels to bring new products to market, and so on.

Because of the high risks involved, the scarcity of expensive expertise, and the relatively small scale required at each of these stages, it is hard for a single company to support the full chain of activities in-house. As a result, the industry is developing a series of multilevel consortia in which each enterprise – and each new product introduced – has its own network of contract relationships with research, clinical, production, and marketing groups around the world.

The semiconductor and electronics industries are becoming similarly structured, with independent design, foundry, packaging, assembly, industrial distribution, kitting, configuration, system analysis, networking, and value-added distributor groups essentially acting as service units that perform specialized tasks for one another. Because of services' high value added and extremely low shipping costs, the most powerful consortia now seek and coordinate the world's best suppliers on a truly global basis.

In another set of major restructurings, large service companies are exploiting their technological power and size in ways that have erased many traditional industry boundaries and distribution patterns. This process is most apparent in banking and financial services, where specialists of all sorts have captured important segments of familiar value chains and cross-competition is rampant. But this process is very much alive in other service and manufacturing sectors.

> U.S. industry can regain strategic advantage through the service sector.

For example, retailing, wholesaling, and manufacturing companies now partner and cross-compete on many levels. Many large retailers like Sears, K mart, and Wal-Mart design or produce their own products worldwide, run their own distribution operations, and compete directly with wholesalers, drugstores, and supermarkets. Wholesalers like McKesson are also large manufacturers of private brands, and they support extensive independent retail networks with electronic accounting, billing, merchandising, planning, and control systems. Broad-line manufacturers like Matsushita control thousands of wholesale and retail outlets, which they use both to gather better market information and to assure rapid market access for new products.

Similarly, boundaries between transportation, communications, and travel-service industries are disappearing as airlines begin to provide direct reservations, tours, conferences, auto and hotel arrangements, in-flight telephone service, electronic retailing, and package delivery services in competition–and coalition–with thousands of other service units. Accounting, software, and professional service firms are also breaking all traditional industry boundaries as they openly compete and cooperate to develop and supply new and better products and services to each other and to a wide variety of common customers.

The bottom line is simple but powerful: whenever a company outsources, sets up a joint venture, or enters into a strategic partnership with a major service company–as most will–it is likely to encounter a totally new set of potential competitors and competitive interests. Under these circumstances, an activity-focused competitive analysis, which includes all possible cross-competitors, provides a much clearer basis for assessing the opportunities, relationships, and threats these complex new situations offer. Conversely, defining the company's industry and competitors in product rather than in service-activity terms can lead to strategic disasters.

The capacity to command and coordinate services activities, supplier networks, and contract relations across the globe has become perhaps the most important strategic weapon and scale economy for many of today's most successful enterprises. Indeed, our study suggests that one reason large Japanese export manufacturers have been so successful is that they have avoided the controlled vertical integration of U.S. companies. Instead, they have moved directly to a kind of quasi-integration similar to what we are describing here: coordinating many independent suppliers through strong contractual and strategic relationships. U.S. industry, with its more efficient service sector, now has an opportunity to jump to a higher level of coordinated disaggregation and thus to regain its strategic advantage.

Reprint 90212

The Service Factory

by Richard B. Chase and David A. Garvin

The factory of the future is *not* a place where computers, robots, and flexible machines do the drudge work. That is the factory of the present, which, with money and brains, any manufacturing business can build. Of course, any competitor can build one too – which is why it is becoming harder and harder to compete on manufacturing excellence alone. Lower costs, higher quality, and greater product variety are like table stakes in poker – the price

Richard B. Chase is professor and director of the Center for Operations Management Education and Research at the School of Business Administration at the University of Southern California. David A. Garvin is professor of business administration at the Harvard Business School and faculty chairman of the School's executive program, Manufacturing in Corporate Strategy.

that companies pay to enter the game. Most products can be quickly and easily imitated; and the most automated design and production processes cannot decisively beat the second most automated. Who wins and who loses will be determined by how companies play, not simply by the product or process technologies that qualify them to compete.

The manufacturers that thrive into the next generation, then, will compete by bundling services with products, anticipating and responding to a truly comprehensive range of customer needs. Moreover, they will make the factory itself the hub of their efforts to get and hold customers – activities that now are located in separate, often distant, parts of the organization. Production workers and factory managers will be able to forge and sustain new relationships with customers because they will be in direct and continuing contact with them. Manufacturing, in short, will become the cortex of the business. Today's flexible factories will become tomorrow's service factories.

About 200 years ago, when horse-drawn carriages were made largely by craftsmen, the most successful carriage maker was invariably the most accommodating. Though he prided himself on being a technician – a manufacturer – his success depended heavily on his willingness and ability to talk with customers at key points: before the sale, so he could get a clear idea of what the client needed and what features would satisfy him; during the manufacturing process, so he could incorporate any necessary changes in the product; and after delivery, so he could learn what features had worked (and what hadn't) and what the client needed for maintenance, repair, and replacement.

Even the most flexible factory does not yet capitalize on the full range of services manufacturing could provide.

Mass production overtook customized craftsmanship because customers came to value standardized goods over higher priced, personalized goods. As a result, work grew increasingly compartmentalized through the division of labor. Craftsmanship (that is, manufacturing) became separated from downstream activities, like sales and postpurchase service, as well as from upstream activities, like new-product development and design. Gradually, manufacturing received more and more of its information and instructions through filters – divisions and departments that were separated, functionally and physically, from the production site. Not surprisingly, manufacturing managers complained that those who defined their work rarely understood it or cared enough about its details, problems, or technical possibilities.

For decades, companies muddled through. In recent years, as Japanese competition put pressure on manufacturing businesses everywhere, manufacturers have worked mightily and successfully to educate workers and break down some of the barriers between their upstream activities and the work of the factory. They have encouraged interfunctional communication between product designers and manufacturing engineers and between R&D and quality managers on the factory floor.

These imaginative efforts to accelerate product innovation and improve manufacturing performance were necessary and important. But they are no longer adequate. Today downstream activities have to be joined to the tasks of the factory too. Increasingly, factory personnel have the means to support the sales force, service technicians, and consumers. This support should, and will, be used. Competition is shifting away from how companies build their products to how well they serve customers before and after they build them.

Some of America's best-run companies – Hewlett-Packard, Allen-Bradley, Caterpillar, Frito-Lay – already operate factories whose activities reflect the new role of service in manufacturing competition. None of their facilities is a complete service factory. We are still many years from that. But in the range of upstream and downstream activities these factories perform, and in the degree of interaction between production workers and customers, they point the way to the future.

Service for a manufacturing company inescapably revolves around its products – their design, features, durability, repairability, distribution, and ease of installation and use. Even the most traditional factories of yesterday proffered service of a kind, but their conception of service was narrow. To old-guard factory managers, service was little more than a commitment to meeting due dates. Logistics and distribution urged the factory to complete orders in a timely fashion, to give advance notice of delivery problems, and to package materials for ease of shipment and damage control. Customers were simply numbers on a production schedule.

The barriers between factories and upstream product designers have fallen. Break the barriers to downstream customers too.

Managers in flexible factories have broadened the concept of service to include both a commitment to product variety and a capacity to see the customer in specialized terms. The key to flexibility has been the ability to produce small lot sizes efficiently and to change over quickly from one product to another. Yet even the most flexible factories are not capitalizing on the full range of services they could provide. As a result, they are bypassing competitive opportunities.

Factories generate information and skills that are critical for product design. Among them are accurate and timely feedback on the manufacturability of new designs, the ability to construct prototypes quickly, and the capability of introducing engineering change orders smoothly. But factories are also a resource for helping customers with installation, maintenance, and troubleshooting. The people who make products are often more knowledgeable about their performance, variability, and repair than the people in field service.

Such product experts can also contribute decisively to sales and marketing efforts. Not only do factories generate quality information that helps sales, but producers of very sophisticated products – chip makers, for instance – prove the quality of their products by showing customers that their procedures are virtually infallible. Factories can serve as real-time demonstrators of the technology and systems the company sells. They can also help make salespeople more responsive by supplying detailed cost and production data for competitive bidding and cross-selling and by making an ever-increasing range of products. Once the products have been sold, the factory can also participate in after-sales support by providing accurate and easily accessible information about the status of orders and by replenishing critical parts and products without delay.

Building a full-fledged service factory – one that offers all these services – in one fell swoop would be difficult. The way to reach that goal is to start by identifying a focused set of customer needs and then try to fill them with all deliberate speed, beginning with those services the factory is particularly good at. This means coalescing around a single service model that is so explicit and transparent that the whole organization can easily understand what needs to be done and why.

Here we propose four alternatives; there are undoubtedly more.

The Laboratory. The factory's competence in testing new products and processes and in ensuring manufacturability and quality equips it also to serve as a lab. The archetype is the pilot plant. Factory managers make crucial information available to the rest of the business: data on product performance and architecture to R&D; process parameters to designers; capacity restrictions to sales and marketing. To do so, they need processes that are heavily instrumented, equipment that is capable of fine adjustments, operators who are technically skilled, and documentation that is careful and precise.

Who better than factory personnel to help customers learn the uses and maintenance of products?

Allegheny Ludlum, a large manufacturer of specialty steel, has developed a formal system of process experiments, often high-risk experiments governed by special procedures to encourage innovation. High-risk experiments are those that push the frontiers of steel making; more traditional experiments aim to modify or improve existing operations that remain imperfectly understood.

For example, in the mid-1980s, Allegheny Ludlum began a series of experiments to develop a metallic replacement for the ceramic substrate in automotive catalytic converters. The experiments involved a new family of alloys, which are difficult and time-consuming to manufacture. To begin the experiments, R&D or manufacturing people first obtained sign-offs from the senior vice president of the technical division and the vice presidents of technical services, production, and sales; then, operating managers ran the experiments freely, knowing that the results would not appear in their budgets or detract from their margins.

The company supports such experimentation with an extremely detailed cost accounting system – 100,000 standards, tracking costs for every department, machine, process, grade, width, and gauge of steel. As a result, every experiment yields useful cost and productivity information. The system also tracks defect rates and other quality data. Experimentation has produced impressive results: between 1984 and 1988, the productivity of Allegheny's melt shop and continuous caster improved 7% to 8% per year.

The CEO of Chaparral Steel, a highly successful mini-mill, is explicit about this approach. He is trying, he says, to get his company's research and development done right on the factory floor. First-line supervisors enjoy regular sabbaticals, during which they visit other steel mills, explore new technologies, and visit customers. They bring fresh, often experimental ideas back to the factory and take the lead in introducing them to the shop floor.

Laboratories do more, however, than perform experiments. They also supply supporting data based on rigorous tests – like the clinical labs that support physicians. Such information can also alleviate customers' concerns about product failures and flawed performance. Because customers are less patient and more insistent than ever, the factory that's close to them and does its own product experimentation can respond to their concerns quickly and directly by providing hard data with which they can make judgments about product quality, flexibility, and the like.

At Hewlett-Packard's Fort Collins Systems Division, which makes computers and technical workstations, the quality department has played a lead role in generating information crucial to customers. At first, quality managers developed customer surveys and collected field data on competitors' products for their own use. Then they discovered that marketing viewed the initiative as an encroachment. So they suggested forming a new partnership drawing on the factory's special expertise. The quality department embraced the role of a service department for marketing, working with members of the technical marketing staff to support field sales and service personnel. According to Henry Kohoutek, the quality manager: "The focus for selecting opportunities to contribute was shifted to the marketing staff, who knew well where help was needed and would be appreciated."

Over time, new kinds of information were generated, packaged, and presented in novel formats: product-quality data sheets; test conditions and test-result data summarized in easily understandable ways that impressed customers; videotapes documenting actual product tests and field performance. The quality department also cultivated direct contacts with sales and service personnel through training sessions, presentations, and guided tours through its facilities. Marketing now covers nearly all the associated costs from its own budget.

The Consultant. Hewlett-Packard's experience is closely related to another factory service–problem solving out in the field. Many companies sell technical expertise along with the product. Most use so-called "systems engineers" or technical service people. Yet factory workers can constitute a pool of talent that can perform some of these tasks. The more factory people work with design, marketing, and quality, and the greater the sophistication of the machines they run, the more they come to think like product and process engineers. In the words of one product designer we know, they become "experts on what the product is trying to be." Factory people can teach customers quality control techniques. They could also be made a part of a customer's design team, consulting on manufacturability, or could help customers solve production difficulties.

If customers have questions about Tektronix's oscilloscopes, they can call an "800" number that rings on the factory floor.

Raritan River Steel Company, a metal-rod fabricator, sends factory workers out with salespeople to troubleshoot customers' quality problems; after returning to the factory, they develop solutions with coworkers on the shop floor. In one case, a customer's problem was traceable to a subtle manufacturing flaw–minute scratches on the surface of a rod–that only a worker's trained eye could identify.

Tektronix, a manufacturer of electronic equipment, has pioneered direct communication between customers and shop-floor employees. Into the shipping carton of every oscilloscope it sells, the company inserts a postcard listing the names of the workers who built the scope along with an "800" number to a phone on the shop floor. Every day the factory gets several calls from customers; the six people working in the repair area who answer them have all received telephone training.

Customers call for various reasons: questions about the use of their oscilloscopes, complaints about quality performance, requests for information about other Tektronix products, and so on. Workers and managers meet daily to discuss these calls; if necessary, further conversations with the customer follow up the meetings. In some cases, workers will call customers six months after delivery to find out how well their products are performing.

Of course, to be useful consultants, factory workers must know more about their products and markets than they can absorb by daily osmosis. They should get training in basic quality assurance techniques and education in key technical concepts and terminology. They must also be at ease dealing with customers and even perhaps in making formal presentations.

The Showroom. Another, somewhat unexpected way a factory can be used to advantage, especially in high-tech businesses, is as a showroom. The factory can serve as a working demonstration of the systems, processes, and products it manufactures. It can also dramatize the company's manufacturing superiority and, by implication, its superior quality or reliability.

Allen-Bradley, a manufacturer of industrial automation controls, uses its computer-integrated manufacturing operation in Milwaukee in just these ways. The facility is a 150 feet by 300 feet "factory within a factory." Within 24 hours of order placement, it can produce 1,025 different electronic contactors and relays in lot sizes down to one and with zero defects.

Even before this facility was constructed, the company's sales group had developed a "productivity pyramid" to show the linkages among the various levels of factory control (computer controls at higher and lower levels—plan level, center level, and so forth) and had seized on the idea of using the company's own operations to prove the system's plausibility. The Milwaukee plant also shows off Allen-Bradley's software products, and it models what would otherwise be an abstract and elusive systems architecture.

Factory responsiveness is the linchpin of after-sales support.

The factory can also reinforce customer perceptions of product quality. Copeland Corporation, a manufacturer of air-conditioning compressors, introduced a new model in the 1970s and built a state-of-the-art facility equipped with programmable controllers and other advanced automation. The corporation then arranged customer tours to demonstrate the plant's precision manufacturing and exhaustive test procedures and to underline the product's reliability. Copeland seized 25% of the market in two years.

For the factory to be a successful showroom, three things are needed. Marketing and manufacturing personnel have to work closely together to both understand customers' expectations and fulfill them. Manufacturing managers and shop-floor employees have to be well trained in communication and presentation. And the factory's layout must allow for stopping points and audiovisual aids to highlight processes with the greatest sales potential.

At Frito-Lay's Vancouver, Washington plant, for example, entrances and exits have been enlarged to accommodate large groups of visitors, and microphones and loudspeakers have been located at key observation points. Floors have been reconditioned for safety, and public affairs people have developed three different tours, one tailored to wholesalers, one to retailers, and, of course, one to snackers. One of these tours was critical to landing a $2 million account.

The Dispatcher. Finally, the factory can be the linchpin of after-sales support. Service backup is of obvious importance in emerging industries, where the technology is new and products lack a track record. But factories can also be organized to help companies differentiate products in mature businesses. They can provide customers with quick replacements for defective or worn parts and ensure speedy replenishment of stocks to help customers avoid downtime and stock-outs. An unavailable part can idle an expensive machine – indeed, an entire factory.

Effectiveness as a dispatcher requires an unprecedented level of flexibility. The factory must be able to anticipate demand surges and respond immediately to customers' crises.

For decades, Caterpillar has been extraordinarily skilled at parts control, warehousing, and logistics; it promises to have repair parts for its tractors and earth-moving equipment available anywhere in the world within 48 hours. This responsiveness is emphasized more than ever by Caterpillar's sales force. Today Caterpillar often advises customers in the area of logistical support and related services and has an active consulting practice.

What connects factories and customers is strategy, not technology.

In clothing, personal care, and small appliance businesses, which are often subject to fads and shifts in fashion, quick responsiveness to the retail trade is crucial. The Limited, a clothing retailer, electronically links its hundreds of stores to a single, centralized computer at its Columbus, Ohio headquarters and then to its Hong Kong textile mills through a real-time system that allows the mills to begin reproducing hot-selling items at the end of the first week of sales.

Such responsiveness depends on ever-tighter customer linkages, which can be achieved in several ways. In the most innovative approach, sometimes called interorganizational systems, factories supply customers with computer terminals that are linked directly to the factory's order-entry and production-control system. Inland Steel uses interorganizational systems, as does Levi Strauss. Several leading food companies are exploring the possibility of linking their order-entry systems to the computers that compile data from supermarket scanners so that they can bypass the purchasing departments entirely.

Companies can also tighten customer linkages more conventionally, without sophisticated information systems. Usually, this requires interfunctional teams consisting of representatives from sales, logistics, and manufacturing. At Frito-Lay's Vancouver plant, for instance, logistics and sales managers and plant personnel schedule regular (weekly, monthly, quarterly) contacts, in person and by telephone. Joint training events are common. Teams work together to reduce late deliveries, eliminate stale product, resolve breakage problems, and meet special promotion needs. The result has been rapid sales growth and, in 1988, the highest combined service and quality rating of any of Frito-Lay's full-mix plants.

Each of these models – laboratory, consultant, showroom, and dispatcher – shows a distinctive approach to factory service. They also show how services overlap. A factory cannot become a good dispatcher without excellent customer information; it cannot become a good consultant without detailed product and process knowledge. Factories become service factories when their managers and workers understand customers' needs as deeply and directly as they know their own products.

Not all factory managers welcome the changes a service factory requires. Most companies used to seal off their core production technology and discouraged interaction with groups outside the factory. Manufacturing managers sought to maximize efficiency and protect the line from outside disturbances; they buffered themselves by storing inventories in locations that were set off from the rest of the organization and from customers.

The managers of service factories, in contrast, have to work in an open system. They need direct and accessible connections to design, to marketing, and to strategic planning, as well as to customers. Computers and telecommunications can help here, to be sure. Computerized ordering systems, expert systems to manage complex sales, computerized logs for after-sales support, computerized catalogs for replacement parts, 24-hour answering machines to take customer complaints, inexpensive fax machines–all of these speed up communication and break down functional barriers.

But what drives the open systems approach is strategy, not technology–the fact that management sees the competitive advantages of working more collaboratively. At Frito-Lay's Vancouver plant, the best example of an open system we've come across, the technology is nothing more than a telephone.

Tomorrow's leading manufacturing companies will be the ones whose managers unleash the service potential of their factories. Competition demands it. Technology makes it possible. Indeed, the competitive forces driving companies to differentiate products with new and imaginative services are simultaneously empowering factory organizations to deliver them.

The forces that are driving manufacturing to compete on services also empower the factory to deliver them.

Think again about the division of labor, how the factory organization has evolved–and why. Victorian capitalists (and, later, the "scientific management" movement led by Frederick Winslow Taylor) divided human labor into ever more simplified and automatic tasks. The goal was improved productivity: if every worker did a small part of the job, unskilled workers could build products more quickly and efficiently (if less pleasurably) than craftsmen. The point was to compete on price.

The Japanese raised this principle to a higher, more sophisticated standard. Taking their cues from Taylor and his disciples, Japanese managers have analyzed production processes with astonishing precision–movements in microseconds, tolerances in fractions of microns. Originally, the idea was simply to make people work more efficiently. But this attention to detail also laid the groundwork for just-in-time and computerized manufacturing. Japanese factories were organized to deliver low-cost products fast without sacrificing superior quality or flexibility. (You could always buy a luxurious, indestructible car in just the right color. But not off the lot, and not for three months' wages.)

To stay in the game, U.S. manufacturers have had to reform their processes and practices. Many have succeeded. Today Harley-Davidson stands up to Honda because of its embrace of quality improvements and just-in-time production methods, when a few years ago the company was on the brink of financial disaster. Every business now has to master the science of manufacturing–the analysis, subdivision, and control of tightly defined conversion tasks. Otherwise, its factories will remain hopelessly unfit for world competition and for the computer programs that might run and monitor the line.

Moreover, U.S. industry now has to select, train, and retrain workers who use and compete with smart (and not so smart) machines. Today direct labor averages less than 15% of the cost of most manufactured goods; in five years that number is likely to seem as extravagant as 3% defect rates recently did.

As factories employ fewer people, the ones who remain must contribute as no computer or robot can.

As factories employ fewer and fewer people, those who remain need to know how to deal with complex machines, software interfaces, and design problems, and how to track quality and appreciate customers. Increasingly, as Peter Drucker and others have argued, the factory requires knowledge workers who will add value by thinking more like general managers, by contributing, as no computer can, by seeing the production system whole and suggesting fresh ways to enhance products. The result is that factory workers and managers will know more about selected aspects of their products' performance and potential than anyone else in the organization.

Drucker has noted that the imperatives of information-based competition are breaking down barriers within businesses and making functional divisions obsolete. He might have added that divisions among upstream and downstream activities are evaporating too. Marketing, research and development, design, manufacturing, sales, after-sales service–all are folding into one another. The service tasks of business can no longer be separated neatly and sequentially from the work of the factory. If factory personnel are indispensible to the interfunctional teams that generate excellent designs–as they are–how much more essential they must be to a business that competes on service.

Reprint 89402

IDEAS FOR ACTION

Exploit Your Product's Service Life Cycle

This cycle, lagging the product life cycle, needs sharp direction too.

by George W. Potts

"It is the capacity for maintenance which is the best test for the vigor and stamina of a society. Any society can be galvanized for a while to build something, but the will and the skill to keep things in good repair day in, day out are fairly rare."–Eric Hoffer in *Working and Thinking on the Waterfront* (1969).

By Eric Hoffer's measure, the United States is a vigorous culture. Repair services for electronic products alone are estimated to cost more than $60 billion annually and employ 400,000 people. The computer industry by itself generates more than $20 billion a year in maintenance service revenues. When including everything serviced in the private sector–elevators, construction machinery, typewriters, vending machines, autos, machine tools, cameras, airplanes, home appliances, industrial controls, turbines, trucks, and so on–the number grows to almost $200 billion, or about 6% of the GNP.

Service operations, however, do not always enjoy the same depth and quality of managerial controls found in the manufacturing environment. As a result, product management tools are often transferred unaltered to the service side. When a product is maturing in its life cycle, for instance, its service cycle is only beginning to generate steam. If a service manager uses this product life-cycle notion unthinkingly, the result can be an excessive spare parts inventory buildup, a self-defeating service pricing strategy, misallocation of field manpower resources, or a premature shutdown of product improvement programs.

> Some 70% of service income came as computer shipments were on the wane.

In this article I explore a way to look at maintenance control for products that have fairly long lives and require a good deal of service. This service life-cycle approach has been a fixture at Data General Corporation for several years.

The service life cycle covers the installed base of products needing maintenance. The installed base consists of the difference between total shipments and total "decay"–that is, the reduction in numbers still in use caused by product wear and discard, the customer's upgrading or switching to a substitute, or cannibalization of the product for spare parts.

Estimating the decay rate is very similar to calculating depreciation. Once a product is shipped to a customer there are measurable odds, in each year thereafter, that it will remain in the installed base. Data General found through experience that decay probabilities are reasonably stable by product class. By spreading the projected rate of decay of hardware shipped during the years of the product life cycle, Data General's forecasters could construct a good approximation of the total decay that would take place.

The difference between the cumulative shipments and this aggregated decay is in effect the service cycle. As the chart on timing shows, whereas the typical computer may reach its peak in shipments after only two or three years, the service cycle can easily last 15 years. An elevator may have a product cycle of only 10 years but a service cycle of 100 years!

At Data General, the service life cycle falls naturally into four phases:

1. Rapid growth–from the first shipment to the peak in the product cycle.

George W. Potts heads Occam Research Corporation, a microcomputer software development company in Wellesley, Massachusetts. For five years he was director of strategic planning for the service division of Data General Corporation.

2. Transition–from the peak in the product cycle to the peak in the service cycle.

3. Maturity–from the peak in the service cycle to the last shipment.

4. End of life–from the last shipment through the last unit in the installed base.

Almost 70% of service revenues for this product came in the last two phases. Four factors determined this phenomenon: distribution and installation delays, which combined to postpone service revenue flow well beyond that for products; the compounding effect of service price increases; product upgrade purchases, often producing more servicing opportunities and, therefore, more revenue; and certain service options (at premium prices) available later in the cycle.

For generation of service *profits* from installations, the last two phases were even more productive: more than 95% came during that period, which began two years after the product cycle had peaked. A main reason for this, of course, was the high revenue flow in those two phases. (See the chart contrasting the revenue and profits flows.)

Other factors were the front-end loading of many service costs, including repairer training, the product launch, and purchase of spare parts, and the sparse installation density of products (higher fixed costs per unit) in the early part of the service life cycle. In other words, about the time that manufacturing, development, and product marketing are leaving the scene, the service profit annuity is just beginning to kick in.

Often, however, there are forces working against this back-end loading of profitability. They include: (1) higher failure rates toward the end of a product's economic life, particularly for mechanical and electromechanical products; (2) the annual compounding effect of wage hikes for field repair personnel, particularly hikes not offset by service price increases; (3) mismanagement of spare parts acquisition, distribution, or repair; and (4) the loss, through neglect and field-force attrition, of repair skills.

In the following pages I take up the four phases of the service life cycle.

1. Rapid growth

At the outset, of course, expansion of the rate and extension of the length of the product sales surge are everybody's goals. There are many ways to gain an advantage through service at this time.

One way is by being very service-price aggressive against major competitors. The new product is perhaps being evaluated on a cost-of-ownership basis, so a trivial service price disadvantage can turn into a big sales headache. It is generally inadvisable to raise service prices during this phase.

A recently rediscovered wrinkle for many companies is the longer warranty, like Chrysler's seven years or 70,000 miles for its cars. These warranties appear to reduce service costs to zero during the period when discounting rates are the lowest in the cost-of-ownership analysis. (Obviously, the service cost is often buried in the product price anyway.) Such a ploy can also help the corporate image by carrying the implicit promise of high product quality.

During this period of product "infant mortality," ensure on-time distribution of enough spare parts to make prompt repairs. Vigorous stocking of field inventory can help build the reputation of the product and maintain the goodwill of "pioneering" customers.

Be ready to respond to complaints of product design or manufacturing process flaws. An effective early-warning system after the product launch will bring great rewards if it leads to engineering-out of the problems in later versions of the product. Moreover, when the buildup of spare parts reaches heavy proportions later, they will be of much higher quality. In this manner, Data General helped improve its product reliability at an annual compound rate of more than 10%.

The early-warning system may also call for upgrading and streamlining diagnostic procedures. Diagnostic methods for a new product that work well in the lab often do not in the field. A process whereby field input is quickly assimilated into diagnostics can pay huge dividends over the service life cycle.

Another aspect of response to customer complaints is training of field maintenance staff. It is better to err on the upside with expenditures on training (as well as on the high-

A Computer's Product and Service Life Cycles

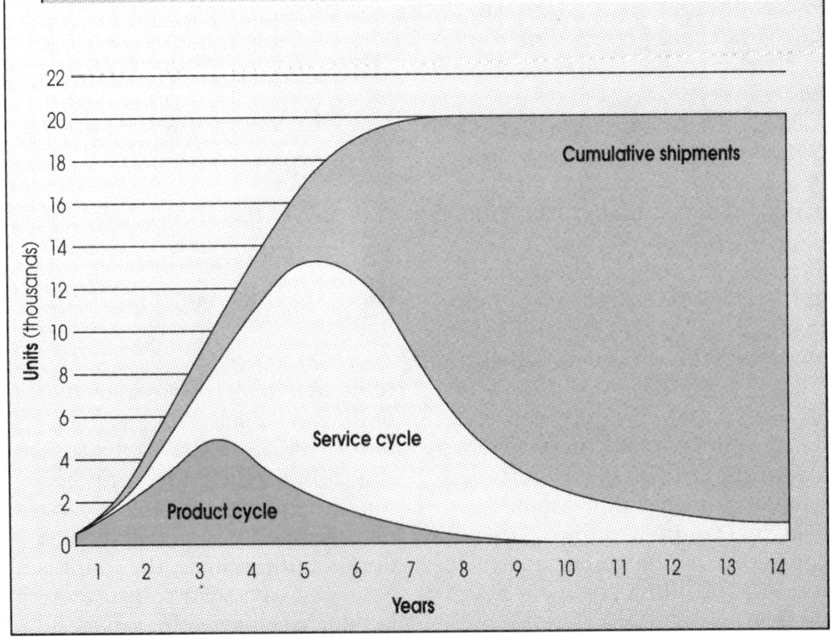

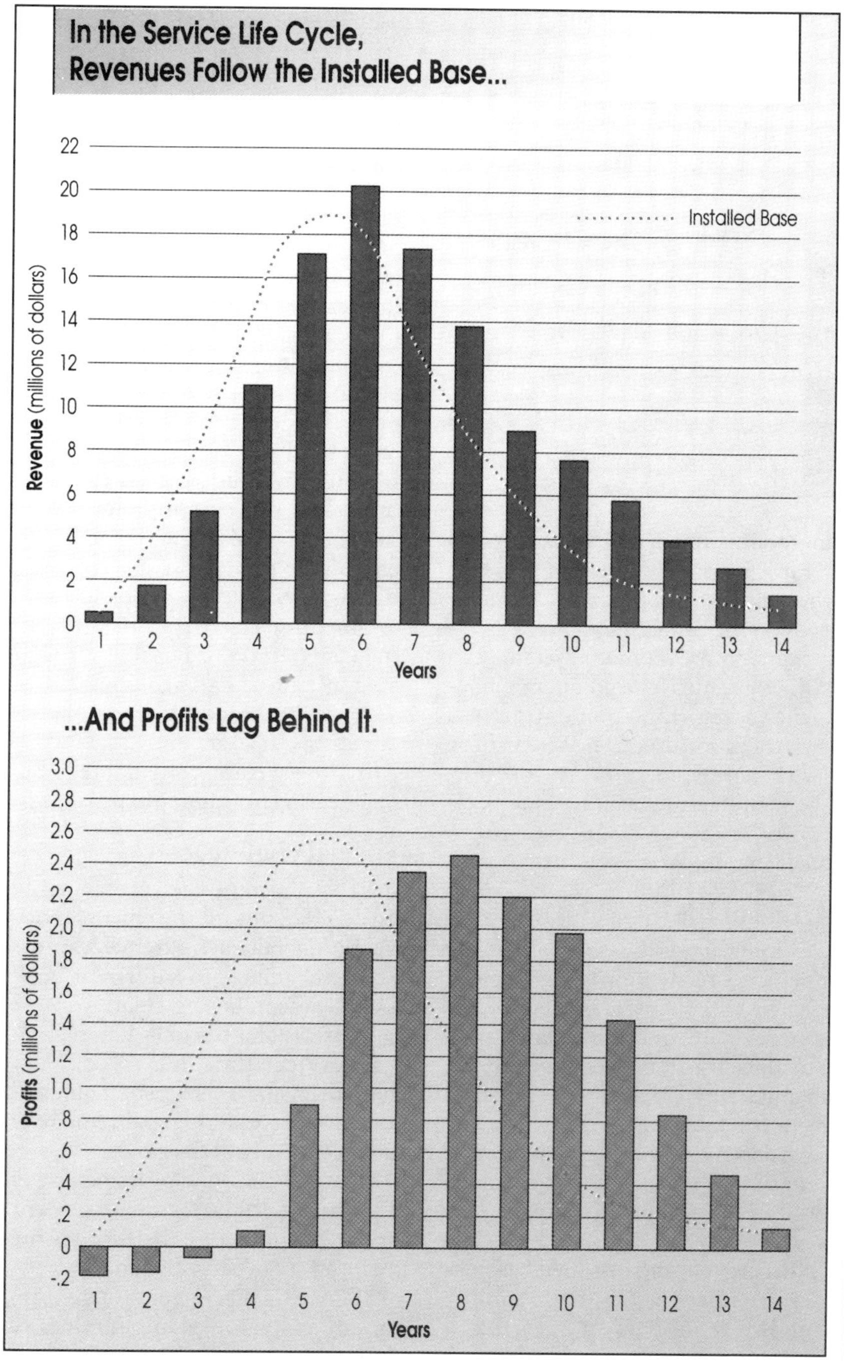

est quality tools and documentation) since continued growth in the residual installed base should easily absorb any apparent lavishness.

Early customers of a new product always appreciate aggressive technical activity of this kind. It alleviates fears of getting a lemon and substantiates their belief that they made the right choice. Satisfied customers may talk to friends and other customers, thus helping the vendor carry out one of the most effective and cheapest marketing techniques known – word of mouth.

2. Transition

During the interval between the time product shipments and service revenue have peaked, profit production from service is usually modest but growing. Whereas the theme of the first phase was "product growth at any cost," the theme of this stage is "controlling service resources." As revenues from the product decline and revenues from maintenance and service rise, management is faced with some critical issues.

One of them is whether to raise service prices. Usually these are not only justified – because the service buildup I described may make the service operation a money loser at this point – but also essential for continued effectiveness. A money-losing service operation is bound to get squeezed in the next budgeting session, which can lead to shortcuts in the initial stage of the next round of products. This, in turn, causes customer defections, which cause further budget cutbacks.

The solution lies in a fair price for service. A fair price often means annual price increases starting early in the transition phase and continuing through the service life cycle. The size of these increments should be tied to the product's failure rate and, to be fair, should seldom exceed the rate of inflation.

If failure rates are declining, it may be strategically smart to cut service prices for profitable products in the later stage of the service cycle. (Companies that don't offer service contracts – and therefore often don't differentiate service pricing by product – can still use the service cycle to set their spare parts pricing strategy.)

The acquisition of spare parts during this phase of the service cycle needs to be closely controlled. It's very easy to look down the road and extrapolate a promise of heavy material usage for a long time. But if the spare parts buildup continues after the growth rate of the service cycle starts to decelerate markedly, serious overstocking of spare parts will be the unhappy result. This will spoil the service profit potential. Therefore, a good fix on the rate of decay is important.

In 1985, field service logistics management at Data General set a policy of restricting purchases of repair material for any product not in phase 1 or 2 of its service cycle. In the three years since, service inventories have been reduced by as much as $30 million.

After the product cycle has peaked, some manufacturers unfortunately fail to maintain the re-engineering activity they carried out in the period of fast sales growth. The service base is still growing, which presents opportunities to reduce product outages, material usage, or repair time. Phase 2 (and even phase 3) is not too late for installing engineering fixes that, though expensive to propagate in the field, still have many years of payback.

Data General has recently concentrated remedial reliability efforts on products no longer in the launch phase. In fiscal 1988 alone, this program is expected to save the service function upwards of $5 million.

Throughout this transition phase, the need for trained repair personnel continues to grow. Since by this time training should have become very efficient and effective, there is little point in being stingy with it now. Training more people than are currently needed may not be a bad idea so that enough maintenance experts will remain after attrition to handle the service requirements throughout the remainder of the product's service cycle.

3. Maturity

As with good wine or cigars, this is the time for the company to reap the rewards of patience. This is harvest time, bringing strong revenue and profit growth. Since all repair asset acquisitions should have been made by now, it will also be a time of very strong cash flow. Moreover, it is a period of the greatest equipment density, with the repair force being at its most proficient.

Unless there is unusually heavy competitive activity, the programmed service price increases should continue. If a product is attracting service competition because its service price is too high, pressure will remain on the manufacturer to make the product more reliable in order to reduce the cost of providing the service. The early part of the maturity phase is a good time to introduce premium service options–car dealers, for example, have a great deal more success selling service contracts on older cars than on new ones–or value-adding options, like the "disaster recovery" services some computer vendors offer.

This is not a time for spare parts aquisition. In fact, every effort should be made to start disposing of material through sale of it, through refurbishing old equipment for sale, through its demanufacture for components, or through scrapping.

The installed base remaining after product sales have declined can yield a service profit annuity.

Some improvement programs can still be pursued for a product that has either a very high population or a very high failure rate. When these two circumstances occur together, the manufacturer still has an opportunity to recoup its investment in such an improvement effort.

4. End of life

While this term suggests that the product is ready for the glue factory, there is still plenty of life left for service revenue and profit generation. But since the item is no longer being manufactured, it may have vanished from headquarters' sight. Nevertheless, depending on the product cycle length and decay rate, as much as 50% of the equipment shipped may still be in use.

This remaining installed base not only yields a service profit annuity but also represents an opportunity for upgrading to newer products. If the manufacturer has prepared a migration path and has established a good service track record, customers may be quite ready to buy the next-generation product.

If service (or spare parts) price increases stopped in phase 3, a return of high failure rates at the end-of-life phase can justify a resumption, particularly for mechanical or electromechanical goods. (High failure rates at the front and back ends of the service cycle are often called the "bathtub effect.") Contract or spare parts price hikes made to offset poor service productivity, balance a defecting customer base, or support unprofitable manufacturing will backfire. But price rises designed to restore a reasonable return to a well-run maintenance function delivering high-quality services are not gouging. Such value pricing is in the best interest of the customer as well as the vendor.

If these hikes encourage upgrading to the manufacturer's newer products, then the service revenue and profit sacrifices become only temporary. If, however, product development has not established a clear migration choice for replacing an end-of-life product, a service price hike is obviously not a good idea.

Continued attrition of replacement material is advisable through scrapping and salvage operations. The only technical programs of benefit at this stage are those that cut repair time without any appreciable engineering or training expense.

Special challenges

Each phase of the service cycle poses problems for the entire service operation, but each extends its most severe challenge to a different business segment. In particular:

Phase 1 (and before)–The technical services unit must work very hard to ensure a successful launch of the product and the rapid remedy of any technical glitches in it.

Phase 2–To avoid overstocking of inventory, logistics management must accurately call the turning point in spare parts demand.

Phase 3–While generating premium and value-adding services to boost total revenue, the service marketing function has to assess the advisability of continued service price increases.

Phase 4–Though the product may be passé in the minds of many at the company, field operations has to understand that it is not so in the eyes of customers. This function has to maintain a staff skilled with the product despite the attrition and relocation of personnel.

Reprint 88512

David A. Garvin

Competing on the eight dimensions of quality

U.S. managers know that they have to improve the quality of their products because, alas, U.S. consumers have told them so. A survey in 1981 reported that nearly 50% of U.S. consumers believed that the quality of U.S. products had dropped during the previous five years; more recent surveys have found that a quarter of consumers are "not at all" confident that U.S. industry can be depended on to deliver reliable products. Many companies have tried to upgrade their quality, adopting programs that have been staples of the quality movement for a generation: cost of quality calculations, interfunctional teams, reliability engineering, or statistical quality control. Few companies, however, have learned to *compete* on quality. Why?

U.S. consumers doubt that U.S. companies can deliver quality.

Part of the problem, of course, is that until Japanese and European competition intensified, not many companies seriously tried to make quality programs work even as they implemented them. But even if companies *had* implemented the traditional principles of quality control more rigorously, it is doubtful that U.S. consumers would be satisfied today. In my view, most of those principles were narrow in scope; they were designed as purely defensive measures to preempt failures or eliminate "defects." What managers need now is an aggressive strategy to gain and hold markets, with high quality as a competitive linchpin.

David A. Garvin is an associate professor of business administration at the Harvard Business School. He has published numerous articles on quality in HBR and other journals and is the recipient of McKinsey Awards for best HBR article in 1982 and 1983. This article draws from his book, Managing Quality, *to be published by Free Press.*

Quality control

To get a better grasp of the defensive character of traditional quality control, we should understand what the quality movement in the United States has achieved so far. How much expense on quality was tolerable? How much "quality" was enough? In 1951, Joseph Juran tackled these questions in the first edition of his *Quality Control Handbook*, a publication that became the quality movement's bible. Juran observed that quality could be understood in terms of avoidable and unavoidable costs: the former resulted from defects and product failures like scrapped materials or labor hours required for rework, repair, and complaint processing; the latter were associated with prevention, i.e., inspection, sampling, sorting, and other quality control initiatives. Juran regarded failure costs as "gold in the mine" because they could be reduced sharply by investing in quality improvement. He estimated that avoidable quality losses typically ranged from $500 to $1,000 per productive operator per year—big money back in the 1950s.

Reading Juran's book, executives inferred roughly how much to invest in quality improvement: expenditures on prevention were justified if they were lower than the costs of product failure. A corollary principle was that decisions made early in the production chain (e.g., when engineers first sketched out a product's design) have implications for the level of

"I spoke to my attorney today, Wendell, and I'm thinking of putting you into play."

quality costs incurred later, both in the factory and the field.

In 1956, Armand Feigenbaum took Juran's ideas a step further by proposing "total quality control" (TQC). Companies would never make high-quality products, he argued, if the manufacturing department were forced to pursue quality in isolation. TQC called for "interfunctional teams" from marketing, engineering, purchasing, and manufacturing. These teams would share responsibility for all phases of design and manufacturing and would disband only when they had placed a product in the hands of a satisfied customer–who remained satisfied.

Feigenbaum noted that all new products moved through three stages of activity: design control, incoming material control, and product or shop-floor control. This was a step in the right direction. But Feigenbaum did not really consider how quality was first of all a strategic question for any business; how, for instance, quality might govern the development of a design and the choice of features or options. Rather, design control meant for Feigenbaum mainly preproduction assessments of a new design's manufacturability, or that projected manufacturing techniques should be debugged through pilot runs. Materials control included vendor evaluations and incoming inspection procedures.

In TQC, quality was a kind of burden to be shared–no single department shouldered all the responsibility. Top management was ultimately accountable for the effectiveness of the system; Feigenbaum, like Juran, proposed careful reporting of the costs of quality to senior executives in order to ensure their commitment. The two also stressed statistical approaches to quality, including process control charts that set limits to acceptable variations in key variables affecting a product's production. They endorsed sampling procedures that allowed managers to draw inferences about the quality of entire batches of products from the condition of items in a small, randomly selected sample.

Despite their attention to these techniques, Juran, Feigenbaum, and other experts like W. Edwards Deming were trying to get managers to see beyond purely statistical controls on quality. Meanwhile, another branch of the quality movement emerged, relying even more heavily on probability theory and statistics. This was "reliability engineering," which originated in the aerospace and electronics industries.

In 1950, only one-third of the U.S. Navy's electronic devices worked properly. A subsequent study by the Rand Corporation estimated that every vacuum tube the military used had to be backed by nine others in warehouses or on order. Reliability engineering addressed these problems by adapting the laws of probability to the challenge of predicting equipment stress.

Reliability engineering measures led to:

Techniques for reducing failure rates while products were still in the design stage.

Failure mode and effect analysis, which systematically reviewed how alternative designs could fail.

Individual component analysis, which computed the failure probability of key components and aimed to eliminate or strengthen the weakest links.

Derating, which required that parts be used below their specified stress levels.

Redundancy, which called for a parallel system to back up an important component or subsystem in case it failed.

Naturally, an effective reliability program required managers to monitor field failures closely to give company engineers the information needed to plan new designs. Effective field failure reporting also demanded the development of systems of data collection, including return of failed parts to the laboratory for testing and analysis.

Now, the proponents of all these approaches to quality control might well have denied that their views of quality were purely defensive. But what else was implied by the solutions they stressed–material controls, outgoing batch inspections, stress tests? Perhaps the best way to see the implications of their logic is in traditional quality control's most extreme form, a program called "Zero Defects." No other program defined quality so stringently as an absence of failures–and no wonder, since it emerged from the defense industries where the product was a missile whose flawless operation was, for obvious reasons, imperative.

In 1961, the Martin Company was building Pershing missiles for the U.S. Army. The design of the missile was sound, but Martin found that it could maintain high quality only through a massive program of inspection. It decided to offer workers incentives to lower the defect rate, and in December 1961, delivered a Pershing missile to Cape Canaveral with "zero discrepancies." Buoyed by this success, Martin's general manager in Orlando, Florida accepted a challenge, issued by the U.S. Army's missile command, to deliver the first field Pershing one month ahead of schedule. But he went even further. He promised that the missile would be perfect, with no hardware problems or document errors, and that all equipment would be fully operational 10 days after delivery (the norm was 90 days or more).

Quality means pleasing consumers, not just protecting them from annoyances.

Two months of feverish activity followed; Martin asked all employees to contribute to building the missile exactly right the first time since there would be virtually no time for the usual inspections. Management worked hard to maintain enthusiasm on the plant floor. In February 1962, Martin delivered on time a perfect missile that was fully operational in less than 24 hours.

This experience was eye-opening for both Martin and the rest of the aerospace industry. After careful review, management concluded that, in effect, its own changed attitude had assured the project's success. In the words of one close observer: "The one time management demanded perfection, it happened!"[1]

Martin management thereafter told employees that the only acceptable quality standard was "zero defects." It instilled this principle in the work force through training, special events, and by posting quality results. It set goals for workers and put great effort into giving each worker positive criticism. Formal techniques for problem solving, however, remained limited. For the most part, the program focused on motivation–on changing the attitudes of employees.

Strategic quality management

On the whole, U.S. corporations did not keep pace with quality control innovations the way a number of overseas competitors did. Particularly after World War II, U.S. corporations expanded rapidly and many became complacent. Managers knew that consumers wouldn't drive a VW Beetle, indestructible as it was, if they could afford a fancier car–even if this meant more visits to the repair shop.

But if U.S. car manufacturers *had* gotten their products to outlast Beetles, U.S. quality managers still would not have been prepared for Toyota Corollas –or Sony televisions. Indeed, there was nothing in the principles of quality control to disabuse them of the idea that quality was merely something that could hurt a company if ignored; that added quality was the designer's business–a matter, perhaps, of chrome and push buttons.

The beginnings of strategic quality management cannot be dated precisely because no single book or article marks its inception. But even more than in consumer electronics and cars, the volatile market in semiconductors provides a telling example of change. In March 1980, Richard W. Anderson, general manager of Hewlett-Packard's Data Systems Division, reported that after testing 300,000 16K RAM chips from three U.S. and three Japanese manufacturers, Hewlett-Packard had discovered wide disparities in quality. At incoming inspection, the Japanese chips had a failure rate of zero; the comparable rate for the three U.S. manufacturers was between 11 and 19 failures per 1,000. After 1,000 hours of use, the failure rate of the Japanese chips was between 1 and 2 per 1,000; usable U.S. chips failed up to 27 times per thousand.

Several U.S. semiconductor companies reacted to the news impulsively, complaining that the Japanese were sending only their best components to

1 James F. Halpin, *Zero Defects* (New York: McGraw-Hill, 1966), p. 15.

the all-important U.S. market. Others disputed the basic data. The most perceptive market analysts, however, noted how differences in quality coincided with the rapid ascendancy of Japanese chip manufacturers. In a few years the Japanese had gone from a standing start to significant market shares in both the 16K and 64K chip markets. Their message–intentional or not–was that quality could be a potent strategic weapon.

U.S. semiconductor manufacturers got the message. In 16K chips the quality gap soon closed. And in industries as diverse as machine tools and radial tires, each of which had seen its position erode in the face of Japanese competition, there has been a new seriousness about quality too. But how to translate seriousness into action? Managers who are now determined to compete on quality have been thrown back on the old questions: How much quality is enough? What does it take to look at quality from the customer's vantage point? These are still hard questions today.

Some consumer preferences should be treated as absolute performance standards.

To achieve quality gains, I believe, managers need a new way of thinking, a conceptual bridge to the consumer's vantage point. Obviously, market studies acquire a new importance in this context, as does a careful review of competitors' products. One thing is certain: high quality means pleasing consumers, not just protecting them from annoyances. Product designers, in turn, should shift their attention from prices at the time of purchase to life cycle costs that include expenditures on service and maintenance–the customer's total costs. Even consumer complaints play a new role because they provide a valuable source of product information.

But managers have to take a more preliminary step–a crucial one, however obvious it may appear. They must first develop a clear vocabulary with which to discuss quality as *strategy*. They must break down the word quality into manageable parts. Only then can they define the quality niches in which to compete.

I propose eight critical dimensions or categories of quality that can serve as a framework for strategic analysis: performance, features, reliability, conformance, durability, serviceability, aesthetics, and perceived quality.[2] Some of these are always mutually reinforcing; some are not. A product or service can rank high on one dimension of quality and low on another–indeed, an improvement in one may be achieved only at the expense of another. It is precisely this interplay that makes strategic quality management possible; the challenge to managers is to compete on selected dimensions.

1 Performance

Of course, performance refers to a product's primary operating characteristics. For an automobile, performance would include traits like acceleration, handling, cruising speed, and comfort; for a television set, performance means sound and picture clarity, color, and the ability to receive distant stations. In service businesses–say, fast food and airlines–performance often means prompt service.

Because this dimension of quality involves measurable attributes, brands can usually be ranked objectively on individual aspects of performance. Overall performance rankings, however, are more difficult to develop, especially when they involve benefits that not every consumer needs. A power shovel with a capacity of 100 cubic yards per hour will "outperform" one with a capacity of 10 cubic yards per hour. Suppose, however, that the two shovels possessed the identical capacity–60 cubic yards per hour–but achieved it differently: one with a 1-cubic-yard bucket operating at 60 cycles per hour, the other with a 2-cubic-yard bucket operating at 30 cycles per hour. The capacities of the shovels would then be the same, but the shovel with the larger bucket could handle massive boulders while the shovel with the smaller bucket could perform precision work. The "superior performer" depends entirely on the task.

Some cosmetics wearers judge quality by a product's resistance to smudging; others, with more sensitive skin, assess it by how well it leaves skin irritation-free. A 100-watt light bulb provides greater candlepower than a 60-watt bulb, yet few customers would regard the difference as a measure of quality. The bulbs simply belong to different performance classes. So the question of whether performance differences are quality differences may depend on circumstantial preferences–but preferences based on functional requirements, not taste.

Some performance standards *are* based on subjective preferences, but the preferences are so universal that they have the force of an objective standard. The quietness of an automobile's ride is usually viewed as a direct reflection of its quality. Some people like a dimmer room, but who wants a noisy car?

2 This framework first appeared, in a preliminary form, in my article "What Does 'Product Quality' Really Mean?" *Sloan Management Review*, Fall 1984.

2 Features

Similar thinking can be applied to features, a second dimension of quality that is often a secondary aspect of performance. Features are the "bells and whistles" of products and services, those characteristics that supplement their basic functioning. Examples include free drinks on a plane, permanent-press cycles on a washing machine, and automatic tuners on a color television set. The line separating primary performance characteristics from secondary features is often difficult to draw. What is crucial, again, is that features involve objective and measurable attributes; objective individual needs, not prejudices, affect their translation into quality differences.

To many customers, of course, superior quality is less a reflection of the availability of particular features than of the total number of options available. Often, choice is quality: buyers may wish to customize or personalize their purchases. Fidelity Investments and other mutual fund operators have pursued this more "flexible" approach. By offering their clients a wide range of funds covering such diverse fields as health care, technology, and energy – and by then encouraging clients to shift savings among these – they have virtually tailored investment portfolios.

Employing the latest in flexible manufacturing technology, Allen-Bradley customizes starter motors for its buyers without having to price its products prohibitively. Fine furniture stores offer their customers countless variations in fabric and color. Such strategies impose heavy demands on operating managers; they are an aspect of quality likely to grow in importance with the perfection of flexible manufacturing technology.

3 Reliability

This dimension reflects the probability of a product malfunctioning or failing within a specified time period. Among the most common measures of reliability are the mean time to first failure, the mean time between failures, and the failure rate per unit time. Because these measures require a product to be in use for a specified period, they are more relevant to durable goods than to products and services that are consumed instantly.

Reliability normally becomes more important to consumers as downtime and maintenance become more expensive. Farmers, for example, are especially sensitive to downtime during the short harvest season. Reliable equipment can mean the difference between a good year and spoiled crops. But consumers in other markets are more attuned than ever to product reliability too. Computers and copying machines certainly compete on this basis. And recent market research shows that, especially for young women, reliability has become an automobile's most desired attribute. Nor is the government, our biggest single consumer, immune. After seeing its expenditures for major weapons repair jump from $7.4 billion in fiscal year 1980 to $14.9 billion in fiscal year 1985, the Department of Defense has begun cracking down on contractors whose weapons fail frequently in the field.

4 Conformance

A related dimension of quality is conformance, or the degree to which a product's design and operating characteristics meet established standards. This dimension owes the most to the traditional approaches to quality pioneered by experts like Juran.

All products and services involve specifications of some sort. When new designs or models are developed, dimensions are set for parts and purity standards for materials. These specifications are normally expressed as a target or "center"; deviance from the center is permitted within a specified range. Because this approach to conformance equates good quality with operating inside a tolerance band, there is little interest in whether specifications have been met exactly. For the most part, dispersion within specification limits is ignored.

One drawback of this approach is the problem of "tolerance stack-up": when two or more parts are to be fit together, the size of their tolerances often determines how well they will match. Should one part fall at a lower limit of its specification, and a matching part at its upper limit, a tight fit is unlikely. Even if the parts are rated acceptable initially, the link between them is likely to wear more quickly than one made from parts whose dimensions have been centered more exactly.

To address this problem, a more imaginative approach to conformance has emerged. It is closely associated with Japanese manufacturers and the work of Genichi Taguchi, a prizewinning Japanese statistician. Taguchi begins with the idea of "loss function," a measure of losses from the time a product is shipped. (These losses include warranty costs, nonrepeating customers, and other problems resulting from performance failure.) Taguchi then compares such losses to two alternative approaches to quality: on the one hand, simple conformance to specifications, and on the other, a measure of the degree to which parts or products diverge from the ideal target or center.

He demonstrates that "tolerance stack-up" will be worse – more costly – when the dimensions

Exhibit **Two approaches to conformance**

In the following graphs, shaded areas under the curves indicate items whose measurements meet specifications. White areas indicate items not meeting specifications.

Production process 1

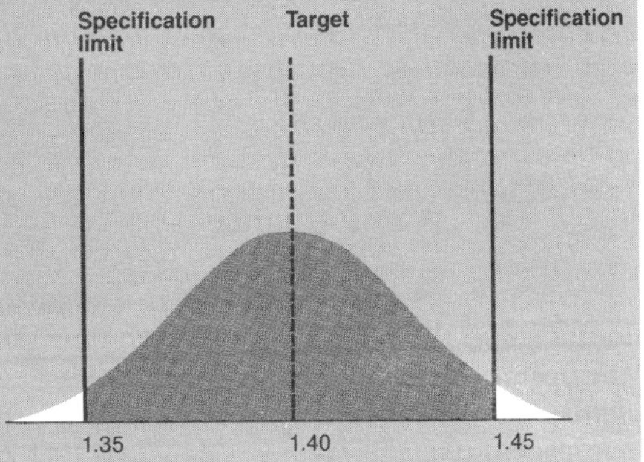

In production process 1 (favored by Taguchi), items distribute closely around the target, although some items fall outside specifications.

Production process 2

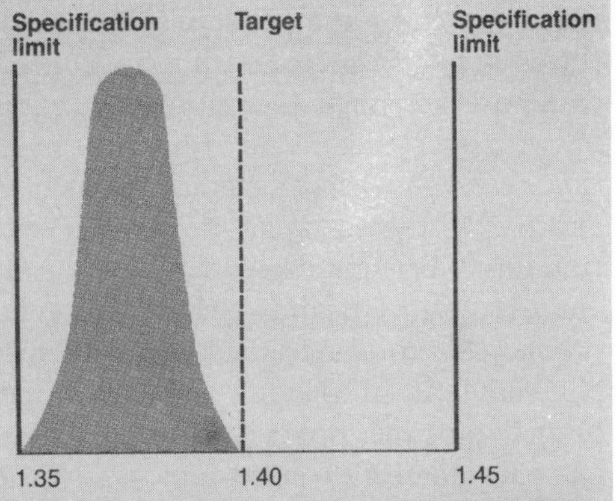

In production process 2 (favored in traditional approaches), items all distribute within specifications, but not tightly around the target.

Source: L.P. Sullivan, "Reducing Variability: A New Approach to Quality," *Quality Progress*, July 1984, p. 16.

of parts are more distant from the center than when they cluster around it, even if some parts fall outside the tolerance band entirely. According to Taguchi's approach, production process 1 in the *Exhibit* is better even though some items fall beyond specification limits. Traditional approaches favor production process 2. The challenge for quality managers is obvious.

Incidentally, the two most common measures of failure in conformance – for Taguchi and everyone else – are defect rates in the factory and, once a product is in the hands of the customer, the incidence of service calls. But these measures neglect other deviations from standard, like misspelled labels or shoddy construction, that do not lead to service or repair. In service businesses, measures of conformance normally focus on accuracy and timeliness and include counts of processing errors, unanticipated delays, and other frequent mistakes.

5 Durability

A measure of product life, durability has both economic and technical dimensions. Technically, durability can be defined as the amount of use one gets from a product before it deteriorates. After so many hours of use, the filament of a light bulb burns up and the bulb must be replaced. Repair is impossible. Economists call such products "one-hoss shays" (after the carriage in the Oliver Wendell Holmes poem that was designed by the deacon to last a hundred years, and whose parts broke down simultaneously at the end of the century).

In other cases, consumers must weigh the expected cost, in both dollars and personal inconvenience, of future repairs against the investment and operating expenses of a newer, more reliable model. Durability, then, may be defined as the amount of use one gets from a product before it breaks down and replacement is preferable to continued repair.

This approach to durability has two important implications. First, it suggests that durability and reliability are closely linked. A product that often fails is likely to be scrapped earlier than one that is more reliable; repair costs will be correspondingly higher and the purchase of a competitive brand will look that much more desirable. Because of this linkage, companies sometimes try to reassure customers by offering lifetime guarantees on their products, as 3M has done with its videocassettes. Second, this approach implies that durability figures should be interpreted with care. An increase in product life may not be the result of technical improvements or the use of longer-lived materials. Rather, the underlying economic environment simply may have changed.

For example, the expected life of an automobile rose during the last decade – it now averages 14 years – mainly because rising gasoline prices and a weak economy reduced the average number of miles driven per year. Still, durability varies widely among brands. In 1981, estimated product lives for major home appliances ranged from 9.9 years (Westinghouse) to 13.2 years (Frigidaire) for refrigerators, 5.8 years (Gibson) to 18 years (Maytag) for clothes washers, 6.6 years (Montgomery Ward) to 13.5 years (Maytag) for dryers, and 6 years (Sears) to 17 years (Kirby) for vacuum cleaners.[3] This wide dispersion suggests that durability is a potentially fertile area for further quality differentiation.

6 Serviceability

A sixth dimension of quality is serviceability, or the speed, courtesy, competence, and ease of repair. Consumers are concerned not only about a product breaking down but also about the time before service is restored, the timeliness with which service appointments are kept, the nature of dealings with service personnel, and the frequency with which service calls or repairs fail to correct outstanding problems. In those cases where problems are not immediately resolved and complaints are filed, a company's complaint-handling procedures are also likely to affect customers' ultimate evaluation of product and service quality.

Some of these variables reflect differing personal standards of acceptable service. Others can be measured quite objectively. Responsiveness is typically measured by the mean time to repair, while technical competence is reflected in the incidence of multiple service calls required to correct a particular problem. Because most consumers equate rapid repair and reduced downtime with higher quality, these elements of serviceability are less subject to personal interpretation than are those involving evaluations of courtesy or standards of professional behavior.

Even reactions to downtime, however, can be quite complex. In certain environments, rapid response becomes critical only after certain thresholds have been reached. During harvest season, farmers generally accept downtime of one to six hours on harvesting equipment, such as combines, with little resistance. As downtime increases, they become anxious; beyond eight hours of downtime they become frantic and frequently go to great lengths to continue harvesting even if it means purchasing or leasing additional equipment. In markets like this, superior service can be a powerful selling tool. Caterpillar guarantees delivery of repair parts anywhere in the world within 48 hours; a competitor offers the free loan of farm equipment during critical periods should its customers' machines break down.

Customers may remain dissatisfied even after completion of repairs. How these complaints are handled is important to a company's reputation for quality and service. Eventually, profitability is likely to be affected as well. A 1976 consumer survey found that among households that initiated complaints to resolve problems, more than 40% were not satisfied with the results. Understandably, the degree of satisfaction with complaint resolution closely correlated with consumers' willingness to repurchase the offending brands.[4]

Companies differ widely in their approaches to complaint handling and in the importance they attach to this element of serviceability. Some do their best to resolve complaints; others use legal gimmicks, the silent treatment, and similar ploys to rebuff dissatisfied customers. Recently, General Electric, Pillsbury, Procter & Gamble, Polaroid, Whirlpool, Johnson & Johnson, and other companies have sought to preempt consumer dissatisfaction by installing toll-free telephone hot lines to their customer relations departments.

7 Aesthetics

The final two dimensions of quality are the most subjective. Aesthetics—how a product looks, feels, sounds, tastes, or smells—is clearly a matter of personal judgment and a reflection of individual preference. Nevertheless, there appear to be some patterns in consumers' rankings of products on the basis of taste. A recent study of quality in 33 food categories, for example, found that high quality was most often associated with "rich and full flavor, tastes natural, tastes fresh, good aroma, and looks appetizing."[5]

The aesthetics dimension differs from subjective criteria pertaining to "performance"—the quiet car engine, say—in that aesthetic choices are not nearly universal. Not all people prefer "rich and full" flavor or even agree on what it means. Companies therefore have to search for a niche. On this dimension of quality, it is impossible to please everyone.

8 Perceived quality

Consumers do not always have complete information about a product's or service's attributes; indirect measures may be their only basis for comparing brands. A product's durability, for example, can seldom be observed directly; it usually must be inferred from various tangible and intangible aspects of the product. In such circumstances, images, advertising, and brand names—inferences about quality rather than the reality itself—can be critical. For this reason, both Honda—which makes cars in Marysville, Ohio—and Sony—which builds color televisions in San Diego—have been reluctant to publicize that their products are "made in America."

Reputation is the primary stuff of perceived quality. Its power comes from an unstated anal-

3 Roger B. Yepsen, Jr., ed., *The Durability Factor* (Emmaus, Penn: Rodale Press, 1982), p. 190.

4 TARP, *Consumer Complaint Handling in America: Final Report* (Springfield, Va.: National Technical Information Service, U.S. Department of Commerce, 1979).

5 P. Greg Bonner and Richard Nelson, "Product Attributes and Perceived Quality: Foods," in *Perceived Quality*, ed. Jacob Jacoby and Jerry C. Olson (Lexington, Mass.: Lexington Books, D.C. Heath, 1985), p. 71.

ogy: that the quality of products today is similar to the quality of products yesterday, or the quality of goods in a new product line is similar to the quality of a company's established products. In the early 1980s, Maytag introduced a new line of dishwashers. Needless to say, salespeople immediately emphasized the product's reliability–not yet proven–because of the reputation of Maytag's clothes washers and dryers.

Competing on quality

This completes the list of the eight dimensions of quality. The most traditional notions–conformance and reliability–remain important, but they are subsumed within a broader strategic framework. A company's first challenge is to use this framework to explore the opportunities it has to distinguish its products from another company's wares.

The quality of an automobile tire may reflect its tread-wear rate, handling, traction in dangerous driving conditions, rolling resistance (i.e., impact on gas mileage), noise levels, resistance to punctures, or appearance. High-quality furniture may be distinguished by its uniform finish, an absence of surface flaws, reinforced frames, comfort, or superior design.

One company's quality niche may be another's trap.

Even the quality of a less tangible product like computer software can be evaluated in multiple dimensions. These dimensions include reliability, ease of maintenance, match with users' needs, integrity (the extent to which unauthorized access can be controlled), and portability (the ease with which a program can be transferred from one hardware or software environment to another).

A company need not pursue all eight dimensions simultaneously. In fact, that is seldom possible unless it intends to charge unreasonably high prices. Technological limitations may impose a further constraint. In some cases, a product or service can be improved in one dimension of quality only if it becomes worse in another. Cray Research, a manufacturer of supercomputers, has faced particularly difficult choices of this sort. According to the company's chairman, if a supercomputer doesn't fail every month or so, it probably wasn't built for maximum speed; in pursuit of higher speed, Cray has deliberately sacrificed reliability.

There are other trade-offs. Consider the following:

- In entering U.S. markets, Japanese manufacturers often emphasize their products' reliability and conformance while downplaying options and features. The superior "fits and finishes" and low repair rates of Japanese cars are well known; less often recognized are their poor safety records and low resistance to corrosion.
- Tandem Computers has based its business on superior reliability. For computer users that find downtime intolerable, like telephone companies and utilities, Tandem has devised a fail-safe system: two processors working in parallel and linked by software that shifts responsibility between the two if an important component or subsystem fails. The result, in an industry already well-known for quality products, has been spectacular corporate growth. In 1984, after less than 10 years in business, Tandem's annual sales topped $500 million.
- Not long ago, New York's Chemical Bank upgraded its services for collecting payments for corporations. Managers had first conducted a user survey indicating that what customers wanted most was rapid response to queries about account status. After it installed a computerized system to answer customers' calls, Chemical, which banking consumers had ranked fourth in quality in the industry, jumped to first.
- In the piano business, Steinway & Sons has long been the quality leader. Its instruments are known for their even voicing (the evenness of character and timbre in each of the 88 notes on the keyboard), the sweetness of their registers, the duration of their tone, their long lives, and even their fine cabinet work. Each piano is built by hand and is distinctive in sound and style. Despite these advantages, Steinway recently has been challenged by Yamaha, a Japanese manufacturer that has built a strong reputation for quality in a relatively short time. Yamaha has done so by emphasizing reliability and conformance, two quality dimensions that are low on Steinway's list.

These examples confirm that companies can pursue a selective quality niche. In fact, they may have no other choice, especially if competitors have established reputations for a certain kind of excellence. Few products rank high on all eight dimensions of quality. Those that do–Cross pens, Rolex watches, Rolls-Royce automobiles–require consumers to pay the cost of skilled workmanship.

6 Consumer Network, Inc., *Brand Quality Perceptions* (Philadelphia: Consumer Network, August 1983), p. 17 and 50-51.

Strategic errors

A final word, not about strategic opportunities, but about the worst strategic mistakes. The first is direct confrontation with an industry's leader. As with Yamaha vs. Steinway, it is far preferable to nullify the leader's advantage in a particular niche while avoiding the risk of retaliation. Moreover, a common error is to introduce dimensions of quality that are unimportant to consumers. When deregulation unlocked the market for residential telephones, a number of manufacturers, including AT&T, assumed that customers equated quality with a wide range of expensive features. They were soon proven wrong. Fancy telephones sold poorly while durable, reliable, and easy-to-operate sets gained large market shares.

Shoddy market research can add quality features nobody wants.

Shoddy market research often results in neglect of quality dimensions that *are* critical to consumers. Using outdated surveys, car companies overlooked how important reliability and conformance were becoming in the 1970s; ironically, these companies failed consumers on the very dimensions that were key targets of traditional approaches to quality control.

It is often a mistake to stick with old quality measures when the external environment has changed. A major telecommunications company had always evaluated its quality by measuring timeliness – the amount of time it took to provide a dial tone, to connect a call, or to be connected to an operator. On these measures it performed well. More sophisticated market surveys, conducted in anticipation of the industry's deregulation, found that consumers were not really concerned about call connection time; consumers assumed that this would be more or less acceptable. They were more concerned with the clarity of transmission and the degree of static on the line. On these measures, the company found it was well behind its competitors.

In an industry like semiconductor manufacturing equipment, Japanese machines generally require less set-up time; they break down less often and have few problems meeting their specified performance levels. These are precisely the traits desired by most buyers. Still, U.S. equipment can *do* more. As one U.S. plant manager put it: "Our equipment is more advanced, but Japanese equipment is more developed."

Quality measures may be inadequate in less obvious ways. Some measures are too limited; they fail to capture aspects of quality that are important for competitive success. Singapore International Airlines, a carrier with a reputation for excellent service, saw its market share decline in the early 1980s. The company dismissed quality problems as the cause of its difficulties because data on service complaints showed steady improvement during the period. Only later, after SIA solicited consumer responses, did managers see the weakness of their former measures. Relative declines in service had indeed been responsible for the loss of market share. Complaint counts had failed to register problems because the proportion of passengers who wrote complaint letters was small – they were primarily Europeans and U.S. citizens rather than Asians, the largest percentage of SIA passengers. SIA also had failed to capture data about its competitors' service improvements.

The pervasiveness of these errors is difficult to determine. Anecdotal evidence suggests that many U.S. companies lack hard data and are thus more vulnerable than they need be. One survey found that 65% of executives thought that consumers could readily name – without help – a good quality brand in a big-ticket category like major home appliances. But when the question was actually posed to consumers, only 16% could name a brand for small appliances, and only 23% for large appliances.[6] Are U.S. executives that ill-informed about consumers' perceptions? The answer is not likely to be reassuring.

Managers have to stop thinking about quality merely as a narrow effort to gain control of the production process, and start thinking more rigorously about consumers' needs and preferences. Quality is not simply a problem to be solved; it is a competitive opportunity.

Reprint 87603

Case Studies

HBR CASE STUDY

Presto Cleaner's new computer system would work great – if it weren't for the customers.

The Case of the Complaining Customer

by Dan Finkelman and Tony Goland

In an effort to improve service, Presto Cleaner installed a new computer system, designed to cut the customers' waiting time and simplify the drop-off and pickup processes. But the system was only a few months old when Mr. J.W. Sewickley, the company president, received an angry letter from Mr. George Shelton, whose laundry had been lost by the new system. Mr. Shelton's letter described his experience with Presto Cleaner's complaint-handling operations and demanded compensation and an apology. To respond to the complaint, Mr. Sewickley sent the letter to his customer complaint office, asking for more information. The answer came back from Paul Hoffner. He explained that there were extenuating circumstances and suggested that some customers may not be worth keeping. Is the customer always right? Where should a company draw the line on compensation and service? What is the best way to handle cases of complaining customers?

HBR's cases are derived from the experiences of real companies and real people. As written, they are hypothetical, and the names used are fictitious.

Dan Finkelman is a principal in the Cleveland office of McKinsey & Company and leader of the firm's worldwide customer satisfaction and service practice. Tony Goland is a senior engagement manager in the Cleveland office.

DRAWINGS BY CHUCK MORRIS

10/24/89

Paul,
This letter came to my attention. Before I can answer it, I need your input. Please advise.
JWS

Mr. J.W. Sewickley
President
Presto Cleaner

October 14, 1989

Dear Mr. Sewickley:

My wife and I are angry, frustrated, and disappointed ex-customers. We weren't always that way. In fact, for a year prior to the recent set of events, we were exceptionally pleased with your service. When you opened your store at the intersection of Adams and Broadway, we were delighted. Even though you're not exactly the least expensive dry cleaner in the area, my wife and I felt that the convenience of the location, the extra early and late hours of operation, and the helpfulness of the staff more than made up for the cost.

That was before you installed your computerized system. The following set of facts will tell you why we are not doing business with Presto Cleaner and what you need to do to get us back as satisfied customers.

July 28: I dropped off some laundry at the store, and the counterperson introduced me to the new computer system. I filled out a "preference card" (light starch for my shirts, folded in a box, and so on) that was entered into the computer. I selected an identification code number (my phone number) and bought a special bag customized with my identification number. The bag was only $3, no big deal. Ideally, the next time I had laundry, all I would have to do was put my laundry in the bag and drop it off. No waiting in line, no waiting for a receipt, the computer knew what I wanted done. When it came time to pick it up, I would just pay, get the laundry, and go. Easy, convenient, time saving. *Supposedly.*

August 4: My wife stopped in to pick up the July 28 order and dropped off the bag with the new laundry (4 of my shirts, 2 blouses, 1 suit, 1 skirt). The counterperson had her fill out her own preference card and entered that information into the computer.

August 10: On the way home from work, I stopped in to make a drop-off and a pickup. Guess what, Mr. Sewickley? I needed to buy a second special bag if I wanted to use the new system every time. I had to stand in line and wait my turn and finally give my order to the person behind the counter. It took forever. The great new system required the counterperson to enter every item, its color or other distinguishing feature, and also what operation I wanted done (clean, press, and so on). It took more than ten minutes. The old system was actually faster.

Then when I finished with the drop-off, I told the counterperson I also had an order to pick up. She asked me for the receipt. I explained that I had used the new computerized system with the bags, so I didn't have a receipt. She asked for my identification number. When she punched it into the computer, it said that my wife had picked up the order earlier in the day.

When I got home, I asked my wife if she had picked up the order. She said she had because she had a business meeting the next day and needed a suit that was at the cleaners. I asked about the bag, and after looking everywhere, including the backseat of the car, we finally determined that she had picked up a previous order and definitely not the order with the special bag. Missing were 4 shirts, 2 blouses, 1 suit, and 1 skirt.

August 11: I called the store from work and explained the problem. The counterperson was very courteous, apologized, and said that the store would be searched for the missing items.

August 14: I went to the store to pick up the last order. I stood in line, waiting for roughly 15 minutes for the two people in front of me to struggle through the computer system. I finally got my order and asked about the lost clothing. After a lot of asking around, the counterperson finally determined that the clothes had not turned up at the store. We next tried to locate the order in the computer, only to discover that when my wife and I had chosen identification numbers, she had used our home phone number and I had used my business number. After searching the computer using both numbers, we still turned up nothing. The counterperson said he would put a tracer on the order back at the plant.

August 15: My wife left work early to take our next laundry order to one of your competitors that has a store near our house. It's open only until 5:30, but there's less nonsense.

August 19: I picked up the order from the nearby Kwik N' Klean on my way to my favorite men's store to buy four new shirts. I didn't have enough shirts to make it through two weeks of work because of the four that were still missing. I am enclosing the sales slip for the four new shirts. I fully expect you to reimburse me for these shirts.

August 22: I called the store to see if my clothes had been found. There was still no word from the plant.

August 25: I called the store again. It had heard from the plant, and the plant did not have the clothing.

Why, Mr. Sewickley, did they not call me? I asked how to pursue a claim for lost items and learned that I should call a Mr. Paul Hoffner at the office.

I immediately called Mr. Hoffner and was told that he was not available. I left a message for him to call back as soon as possible.

August 26: I called back again. Mr. Hoffner was not available. I left the same message.

August 27: I called back again. Mr. Hoffner was not available. I asked if there was anyone else who could handle a claim for lost items and was told that only Mr. Hoffner could do that.

August 31: Mr. Hoffner called. I told him that I wanted to put in a claim. He was totally unaware of the situation, so I had to explain it to him. He suggested that he call the store and the plant to find out if there was any progress. I got the distinct impression that he didn't believe me and that he felt he needed to check with the store to make sure that I really was a customer and that there really was missing laundry. I assured Mr. Hoffner that I was a real customer. In fact, for more than a year, my wife and I have averaged between $20 and $30 worth of dry cleaning every week. Despite my assurances, Mr. Hoffner insisted that he had to check before anything could happen. When I pushed to find out what the process was, Mr. Hoffner said that if the clothing really was lost, I could fill out a claim form and apply for compensation.

September 7: A week had passed with no word from Mr. Hoffner. I called. He wasn't available. I left a message.

September 11: Still no word from Mr. Hoffner. I called and miracle of miracles, he answered the phone. After all that time, the only thing he could tell me was that neither the store nor the plant could find the clothes. I asked him to send the claim form.

September 18: No claim form had yet arrived in the mail, so I called Mr. Hoffner again. I got him and asked about the form. He said he had delayed sending it because he was sure that the clothes would be found. I insisted that he send the form immediately.

September 21: The claim form arrived, and I discovered that it required that I attach both the original purchase receipt for the clothes and the counter receipt for the laundry order. Mr. Sewickley, do you keep your year-old receipts for clothing? And with the new computer system, there are no counter receipts!

It took two more calls to get through to Mr. Hoffner to complain about these requirements. He said that they were necessary to guard the company against fraud. He did acknowledge that since your own system no longer produced counter receipts, I couldn't be expected to send them in. But as far as the receipts for the clothes were concerned, he suggested that I go back through our charge card records to come up with the proof of purchase. I flatly refused. I told Mr. Hoffner that we have done more than $1,000 worth of business with Presto Cleaner over the past year and that we were not in the business of extorting money from dry cleaners. Further, I told him that if my word wasn't good enough for him, he could kiss our business good-bye, along with that of our friends and colleagues at work, who would soon hear all about the Presto Cleaner way of doing business.

September 22: Having had a terribly busy week, we forgot that my wife needed a dress cleaned for a business function Saturday night. Since it was Friday morning when we discovered this, our only real option was for my wife to drop off the dress at Presto Cleaner in the morning.

September 23: I went in to pick up the dress. The woman behind the counter recognized me and told me that the store had found our lost clothes. Apparently, they had been mysteriously included in another customer's order and only just now returned. She had no clue how this could have happened with the new system. She was, as usual, cheerful, apologetic, and polite about the mix-up. I paid for the order, picked up the dress, and went home.

September 25: Since there was no way to call your office over the weekend, I waited until Monday to try to reach Mr. Hoffner. He was again unavailable, so I left him the last message he will ever get from me: the clothes were found, but I would still like to talk to him.

Mr. Sewickley, that was more than two weeks ago, and I still haven't heard from Mr. Hoffner. I am outraged by this entire episode, by the way your company treats customers, by Mr. Hoffner's conduct, by the lack of communication, and by the ridiculous system you introduced. I am particularly incensed at having to pay for clothes that were delivered almost two months late and by having to purchase new clothes to cover your company's mistake.

I expect the following: a full refund for the order that was lost; full payment for the four shirts that I had to buy to make up for the lost order; and a full apology from Mr. Hoffner. If all of those are forthcoming, I might consider giving your company another chance at my business. Otherwise, my wife and I will never patronize your company again.

Sincerely,

George Shelton

George Shelton

Memo
To: JWS
From: Paul Hoffner P. H.
Re: Customer Complaint

Oct. 29, 1989

This is in response to your memo requesting background information to respond to the customer complaint of Mr. George Shelton. I have reviewed his letter as well as our own file concerning this matter. I am convinced that we did make a good-faith effort to do right by Mr. Shelton, although he may not recognize it as such. Nevertheless, should you wish to mollify the customer, I would be perfectly happy to play the role of fall guy if it would help. As far as extending compensation to the customer is concerned, his demands seem to me far in excess of any real liability: he did get his clothes back, he will keep the four new shirts and use them, as well. I would certainly extend an apology to him—if he would like it from me, fine; I assume that it would be even more satisfactory coming from you, along with an offer to clean his next order free of charge.

Having said this, there are some mitigating circumstances that you should be aware of. I would not share these with the customer, but present them to you so that you will understand more accurately and fully what really happened, rather than what this one customer says happened. Again, that does not mean that a mistake was not made; there was a mistake, and we should take responsibility for that. But we acted in a way consistent with company policy and operations. Let me describe what really happened:

1.) The customer dropped off his clothing on literally the first day of operation of the new computerized system—a system that everyone at headquarters agrees is the key to our future success. Everyone also agrees that there will be problems, including, unfortunately, some lost customers. Mr. Shelton may be one such example. As you remember, because of delays with the system vendor and the software, we had only one week to train our people on the new system. Also, as I'm sure you're aware, not all of our store personnel are great at using this technology. It has taken more time than expected to get them to understand all of the steps they have to take to prevent mistakes.

2.) As I suspected all along, the real problem in this case was that the clothing was picked up by another customer. This is not that unusual. When it happens, we must rely on the goodwill of our customers to return what is not theirs. This time, because the customer was an infrequent user of our service, it took him more than four weeks to bring the clothes back in. Also, he dropped the clothes off at a different store. Unfortunately, we have more than one case of lost clothes at a time (please note all the letters of complaint that you *don't* have, proof that the system works most of the time), so we cannot call every customer and ask him or her to come in and identify the clothes. Our policy, therefore, is simply to wait for customers to come in and then offer them a chance to identify the clothes.

3.) Much of Mr. Shelton's gripe concerns the early period (approximately August 12 through August 25) during which time he felt he was kept in the dark. But as you know, our process is designed to get the stores and the plant to figure out problems. My written record shows that the plant did two thorough searches before reporting the lack of results to the store. It's hard to fault it for thoroughness. Apparently, the store also delayed referring the customer to me, probably thinking that the clothes would turn up and, perhaps, not wishing to have to report bad news concerning our new computer system. The process also may have taken this much time because of our standing company policy that requires all reports between stores, the plant, and the office to be in writing.

4.) After I spoke with the customer, I spent the next ten days or so checking with the store and the plant in an attempt to determine what exactly had happened and to locate and identify the clothes. By the end of the first week, I felt sure that the clothes had mistakenly gone to another customer. But there was no way to retrieve them, other than to wait for them to turn up. I did not tell the customer this, of course; after years of experience, I have found that customers only get more upset at the idea that a stranger has their clothes.

5.) Unbeknownst to me, the missing clothes were returned to the Adams and Broadway location on September 14. The customer who had mistakenly received the clothes dropped them off at one of our suburban locations, which then forwarded them to the plant. The plant used our standard identification process, sent them to the Adams and Broadway store, and then wrote me a memo, which I received on September 23. Before I could call the customer and inform him, he had picked up his clothes. I did get one more very angry message from the customer, which I decided not to return, since further communication would unlikely be of any benefit.

6.) Mr. Shelton's letter has several misstatements of fact. While there is no point in disputing these points with him, you should know the following:

(a) I did not delay in sending him a claim form. I sent the form after he requested it and do not know why it was not received. When he called back on September 18, I sent another form. I did not tell him that I had delayed in sending the form. I told him we were trying to locate and identify his clothes.

(b) When I spoke to the customer regarding the standard claim form, I did not tell him to produce proof of purchase. I told him it would be helpful to us if he could get a good estimate of the value of the items.

7.) As you can tell from the tone of Mr. Shelton's letter, he is a very demanding, persistent individual. What his letter does not tell you, however, is that he inundated us with the sheer volume of his calls. Given the number and frequency of his calls, there was no way for me to demonstrate progress on his problem before he called again. When he writes that he "left a message," you should know that his "messages" were usually cryptic. He often did not leave a phone number and, on more than one occasion, even refused to leave his name. For example, he would say, "You know who this is." My secretary found all of this quite distressing. Moreover, I've never run into a customer so anxious to be compensated.

This brings up several interesting questions: How do we make up for the mistake we made without being browbeaten into excess compensation? For example, his four shirts would cost us more than $200; refunding his order would add another $35; if I had swiftly processed his original claim, it would have come to over $600.

In other words, despite Mr. Shelton's threats, I think our system worked. Although we did misplace his clothes, he got everything back, and we avoided a major expense. Now some restitution, such as one free order and a written apology, should be enough. If that is not good enough for Mr. Shelton, it seems to me we should ask, "Aren't there some customers we are better off losing?" Maybe this is a customer that Kwik N' Klean deserves!

How can Presto Cleaner keep Mr. Shelton as a customer? Should it try?

Four authorities on customer service consider Presto Cleaner's customer complaint.

Presto Cleaner's behavior could cost it $500,000 a year.

LEONARD A. SCHLESINGER *is associate professor of business administration at the Harvard Business School.*

Mr. Sewickley should replace Paul Hoffner. His response–and consequently Presto Cleaner's–to the situation appears to be oriented toward driving customers away, not toward building a business.

First, look at the situation. Presto Cleaner should be commended for implementing computer-based systems that increase the speed of service and eliminate paper pushing. However, the decision to introduce the system with limited employee training and no advance customer preparation or education created a nightmare. This could have been avoided. Each operating unit could have invested resources to set up a separate line or section of the store to indoctrinate customers in the new system without disrupting the normal flow of business. Hoffner resorts to identifying the rushed implementation of the system as an acceptable excuse for the customer problem. This clearly is not the case.

Certainly, mistakes will happen–all organizations make them. But mistakes provide organizations like Presto Cleaner the opportunity to recover in ways that actually build customer loyalty rather than lose business. This fact seems to escape Paul Hoffner. He appears to have adopted an internally focused and strictly rational approach to search for the missing clothing. He therefore ignores the customer's emotions. Unreturned phone calls and unsent forms serve only to turn a small mistake into a big one. The assumption that Presto Cleaner must actively guard against the potential of customers' fraudulent claims pervades all aspects of his relationship with Mr. Shelton. It is an attitude that other customers and the Presto Cleaner staff must notice as well. If you regard your customers as a band of potential cheats and believe that you must protect yourself from them, how can you possibly deliver to them the kind of service they deserve?

Having blown several opportunities to impress Mr. Shelton with Presto Cleaner's desire to solve his problem, Mr. Sewickley now faces the question of restitution for the situation. Mr. Shelton is clearly angry, and his request is undoubtedly greater than it would have been a few weeks earlier had Presto Cleaner handled the situation more ably. Paul Hoffner believes the request is absurd and actively encourages Mr. Sewickley to consider losing the Sheltons as customers. But we know from consumer research that for every complaint such as Mr. Shelton's, there are 20 or more customers who leave quietly and unnoticed and thereby provide the organization with no opportunity to learn from its mistakes. Mr. Hoffner is about to ensure that no learning occurs even when the customer complains. Purely in economic terms, the loss could be enormous. Assume that Presto Cleaner manages to lose one customer a day (not an unreasonable assumption given the company's behavior in this situation). Based on Mr. Shelton's estimates of his laundry expenditures, the annual revenue loss from such an unnoticed customer defection would reach almost $500,000.

In that context, Mr. Shelton's request is not at all unreasonable. If I were Mr. Sewickley, I would use the opportunity presented by the complaint letter to make one last attempt at service recovery. I would immediately reimburse Mr. Shelton's $235. But rather than write a simple apology letter, I would deliver the check myself and stress the following points:

☐ What happened was unacceptable; there are no excuses for it.
☐ Presto Cleaner will reimburse all out-of-pocket losses.
☐ Presto Cleaner does not want to lose Mr. Shelton and will do *whatever* it takes to get him back.

My assumption is that this dramatic gesture will bring Mr. Shelton back to Presto Cleaner.

Lastly, Mr. Sewickley should use this complaint and the company's response as a learning opportunity for the company. The organization must establish a clear model of customer relations and service standards at the top. The Shelton complaint provides an opportunity for a broad discussion of customer problem resolution. In a competitive business environment, the risks of not capitalizing on this opportunity could be catastrophic.

Every business should aim to close its customer complaint department.

DINAH NEMEROFF *is corporate director of customer affairs at Citicorp/Citibank.*

Presto Cleaner has a dual challenge: to regain Mr. Shelton's business and ensure consistent customer-satisfying service to all customers. Both issues underscore why this is really "A Case *for* Problem-Free Service." While responsive service after the sale is a prerequisite for organizations today, their primary goal should be problem-free service. Every business should aim to close its customer complaint department! Problem prevention, not problem resolution, is a superordinate service-quality objective.

Unfortunately, Shelton already is knee-deep in problems, and these have grown geometrically, as is often the case. Presto Cleaner's Mr. Hoffner, his interactions with Shelton, and his lack of follow-through have made a bad situation worse. To remedy the situation, Presto Cleaner's president, Mr. Sewickley, should take some immediate first steps. First, he should send a letter of apology to Shelton with two enclosures: a $235 check, to cover the $35 charge for the delayed order and the $200 cost of four new shirts, and a $50 certificate for future Presto Cleaner service. He should act on the principle that, having made service mistakes, Presto Cleaner should bear all the customer's out-of-pocket costs. In addition, to demonstrate that anything short of highly satisfying service is not Presto Cleaner's way, Sewickley should enclose the $50 certificate.

This solution is based on the four-rung customer satisfaction ladder we're working to implement at Citibank. The system works like this: rung 1, customers say "This company doesn't work"; rung 2, customers' basic service requirements are met; rung 3, customers get service personalized to their needs; and rung 4, customers say, "The service wows me!"

Ironically, Presto Cleaner's new computer system was geared to take the company's second rung, basically satisfying service, to a higher level of personalized service–rung 3. However, the total service failure moved Shelton down to rung 1. Therefore, a rung 4, "Presto Cleaner wows me!" action is called for. Sewickley's restorative actions most likely will regain Shelton's business. It should also enhance Presto Cleaner's reputation in the long term.

The larger issue on Sewickley's agenda must be to modernize Presto Cleaner's quality performance, just as he's working to update his technology. Both service and infrastructure should be state-of-the-art. Shelton's experience suggests there is much to do. For example, Presto Cleaner's policies to deal with customer problems require a thorough review. How can the stores and plant respond more quickly to customer difficulties? How can quality be designed into new procedures *before* they're introduced? Such policy reviews, interdepartmental teamwork, and quality-assurance issues are classical service-quality priorities. Sewickley should establish and communicate profitability-based customer satisfaction as a primary company goal and implement a diversified service-quality program.

Two imperatives are key to Presto Cleaner.

Strive for problem-free service. Presto Cleaner needs to "do it right the first time." As Shelton's experience illustrates, customer problems are costly to resolve, lead directly to lost business, and erode customer goodwill.

Even with first-rate customer service, customers do not "fully recover" from service problems. Citibank research shows that customers who've had problems resolved *to their satisfaction* are, nevertheless, on average 20% less satisfied than problem-free customers. Also, Presto Cleaner's management should not make the common mistake of viewing customer complaints as a comprehensive report card of the company's service. Complaints are a small subset of customer problems, registered by the vocal few; the majority of troubled customers silently desert to competitors. Presto Cleaner executives should understand the type, frequency, and severity of customer problems to implement a problem-free service plan. Such experiences must be seen from the customer's perspective. In large-scale organizations – or when employees rarely interact with customers – systematic customer feedback must be used to gather this crucial data.

Improve customer service. Because it's a rare business that can close its customer complaint department, this service function must be efficient and responsive.

Presto Cleaner's customer service problems begin with the customer-unfriendly computer system and Paul Hoffner, but they do not end there. A Presto Cleaner customer with a problem is in a limbo of missing laundry, knowing neither what to expect nor how to get help.

I recommend three strategies we've found valuable at Citibank:

☐ Trust the customer. Service policies that assume customers are honest, and it's only the occasional crook who must be caught, are the appropriate ethical standard.

☐ First-request service. A customer should only have to contact a business once to have his or her problem solved or question answered – ideally, this can be done immediately. If this isn't possible, the business takes on the follow-up burden.

☐ Manage customers' expectations. When follow-up is necessary, telling the customer what to expect can significantly increase his or her satisfaction. When the service investigations department in one Citibank business instituted this particular procedure, specifying time frames for next steps, customers' satisfaction increased by 40%.

Such service improvements are neither easily nor quickly made, in Presto Cleaner or any organization. In fact, they can be expensive to implement, since up-front investment is frequently required. Nevertheless, the long-term payback is rewarding. A business *can* significantly decrease its cost of quality – foremost by preventing problems, not just fixing them. And when a company consistently provides customer-satisfying service, that business can keep and grow its customer base and earn more of each customer's business.

Mr. Sewickley should "buy the problem."

RON ZEMKE *is president of Performance Research Associates in Minneapolis, Minnesota and author of* The Service Edge *(New American Library, 1988).*

There are three rules of thumb for creating an effective customer service focus in an organization: understand and respond to customers' expectations, do whatever it takes to be perceived as "easy to do business with," and be good at problem solving. Presto Cleaner has done an excellent job of ignoring all three.

In our studies, good recovery – fixing things that have gone wrong for the customer – accounts for 38.7% of the differences between companies perceived as very good or very bad at serving customers. This has proven to be an especially important distinguisher in retail operations like Presto Cleaner.

For the long term, Mr. Sewickley is going to have to do four things to ensure that the level of problem solving at Presto Cleaner is at least competitive. First, he'll have to make it clear that customer-contact employees, like Paul Hoffner, are expected to go out of their way to solve customer problems. Second, he'll have to set standards for both response time and appropriateness of redress. Third, he must see that customer-contact people are henceforth trained to deal with customers who have been, or believe that they have been, wronged by the company. And finally, he is going to have to scrap his marvelous new system or get help redesigning it from a customer-use point of view. As it stands, the new system threatens to make Presto Cleaner about as customer friendly as a state department of motor vehicles.

Right now, however, Mr. Sewickley must take over as Presto Cleaner's chief problem solver. It is especially important that he not follow Mr. Hoffner's advice and write off the Sheltons. Through word-of-mouth, the Sheltons can do a great deal of damage to the business. A simple "You won't believe how the dry cleaner messed us around!" tale, told often enough, can cost Presto Cleaner as much as 2% of annual revenues. Beyond that, Hoffner's assumption that the Sheltons are a lost cause is probably incorrect. The fact that they bothered to write to the company – something less than 10% of unhappy customers will do – suggests that they want to be coaxed back into the Presto Cleaner fold.

Here's what Mr. Sewickley should do to make it easy for the Sheltons to continue being Presto Cleaner customers. Call George Shelton and apologize–profusely, if necessary. The goal is to make sure that the customer knows that Sewickley doesn't condone lackadaisical treatment of customers or consider problems such as the one the Sheltons experienced as par for the course. And he should be prepared to listen more than he talks. Mr. Shelton probably has a large amount of pent-up frustration that he wants to dump on someone. So let him dump it. It won't cost a dime to hear him out, and frequently the sum and substance of what an upset customer wants is to be heard. Sewickley should try to enlist Mr. Shelton both in solving the problem his family encountered and in suggesting ways that the new system can be made more user friendly. Until now, Shelton has been denied any sense of control over the situation. Give him back some control. Enlisting upset customers to create acceptable solutions also works to bring them from an upset and irrational mind-set to a thinking and reasonable one.

If necessary, Mr. Sewickley should be prepared to "buy the problem"–spend the $235, the cost of the new shirts and the dry cleaning order. The odds are, however, that Mr. Shelton will drop his demands that Presto Cleaner pay for his replacement shirts once he is talking to the president.

Remember: Sewickley is trying to install a new system. If too many customers are walking around saying bad things about the new way of doing business with Presto Cleaner, there will be no business to do.

And he should follow up with some form of symbolic atonement. He definitely has to send a personalized follow-up letter that refers to their conversation and what will happen as a result of it. But at this point, an additional gesture is necessary. A certificate for free work is OK, flowers would be better.

In the last analysis, Mr. Sewickley is both arbiter and role model for customer relations and customer treatment at Presto Cleaner. If he "blows off" a customer who is initially unreasonable or uses bad language or is simply unpleasant, Mr. Hoffner and every counter clerk in every store will do the same. If Mr. Sewickley shows patience and understanding and evidences a sincere desire to do what it takes to keep Presto Cleaner customers happy and coming back, everyone in his employ will follow suit.

Send Shelton to dinner and Hoffner to the front lines.

CLAUS MØLLER *is president of Time Management International, a management development company in Denmark.*

Presto Cleaner's poor service on the front lines reflects the unresponsive and irresponsible employee attitudes of upper and middle management. Mr. Hoffner is not the villain in this case but the victim of a service-management culture that has not built a support system that encourages employees to demonstrate a positive attitude toward the customer.

Customers who take the time and energy to complain are doing companies a favor. They help companies stay in business. Mr. Hoffner must change his attitude; Presto Cleaner should too. When customers complain to Presto Cleaner, employees should say thank you. This company should train its employees to listen to customers and change fundamental attitudes so that customers' complaints are viewed as opportunities for positive change, not as reasons to be defensive.

Quality management involves changing attitudes at every level of the company. It is not just for managers. Nor is it just a question of implementing technology and systems. The problem in this instance lies not in the details of what happened but in the flawed values embedded in the procedure.

Mr. Shelton's disappointment stems from bad customer service rather than bad technology. Customer dissatisfaction is often based on emotions, but when asked to explain why they are dissatisified, customers give rational explanations–even when they are not the real cause. Let me illustrate. When I worked with SAS years ago to improve the company's service, we asked flight attendants to distribute forms asking customers to rate the service. I observed that the process itself was flawed: attendants did not make eye contact, and when customers asked, the attendants would not explain the reason for the forms.

Not surprisingly, customers responded negatively. And when they did, they commented on what I call hard service quality: material aspects such as the space between the seats, waiting time, punctuality of flights, even the fact that champagne was served in plastic glasses.

I then trained attendants to interact better with customers–letting them know why the forms were being distributed or why planes were not on time. After that, the customers responded to the forms by saying they were satisfied. And they named all the same hard qualities as the reasons for their satisfaction! When people are not aware of the "soft" qualities for their dissatisfaction, they invent a lot of socially "reasonable" explanations as justification.

Likewise, the problem here is not the specific incidents that took place but the attitude Mr. Shelton encountered. Had Mr. Hoffner demonstrated a genuine interest in

Mr. Shelton's problem and given less attention to defending himself, I doubt that the customer would have been dissatisfied. Instead of showing interest in Mr. Shelton's problems, Mr. Hoffner refused to admit to either his customer or his boss that he might be wrong. He is more interested in making the customer wrong than in admitting his own error. He treats the customer as an irritant.

The actions of Mr. Hoffner signal a deeper flaw in the company – which clearly has no defined service policy. Companies should design systems so that employees relate to customers – not technology. Presto Cleaner has flip-flopped its priorities by implementing a computer system that seems to have been designed more for the shop and company than the clients! At least that is the signal sent to its customers. Presto Cleaner also needs to use its time better. If this company spent the same amount of money and time on motivating its staff that it does on writing notes and memos, it would be much better off.

Clearly the company must improve the attitudes of its staff by creating an environment that promotes the best in people. I distinguish between "internal service" – how employees are treated within a company – and "external service" – how employees treat customers. The external service-quality will never exceed internal service-quality. Companies must create responsive attitudes through internal service. Companies must give employees responsibility to make decisions and be prepared for them to make mistakes. They must treat employees as they expect employees to treat customers. One goal to motivate people is to suggest that they treat their customers as they would treat guests in their homes.

I recommend that the company give Mr. Shelton what he requests – along with something extra, such as a complimentary dinner for him and his wife. I also suggest that Mr. Hoffner work on the front line for a while to see how the new system is working and to stay in touch with customers.

Reprint 90304

HBR CASE STUDY

The Case of the Not-So-Supermarket

The service dilemma: Find enough workers, upgrade their skills, and improve productivity – without giving away the store. Are there solutions in these memos?

by Janice McCormick

Janice McCormick is an associate professor at the Harvard Business School, where she teaches a course on employee relations strategies.

Hilltop Stores

Serving Greater Hartford Since 1926

February 16, 1989

To: Margaret Flynn, CEO Hilltop
From: Ed Boyer, VP for store operations

Our contract with Local 413 expires June 18, so we need to start preparing for negotiations (they're scheduled to begin in eight weeks). Now is a good time to go over what's happened in the last three years and to think about what Hilltop needs to do for the future.

You've always said that change is the name of the game in the grocery business. But who could have anticipated the pace of change at Hilltop since the last contract? In 1985, our stores offered 60 to 70 varieties of fresh produce. Today we offer more than 200 varieties—not to mention salad bars, fresh baked goods, fresh seafood, and prepared foods on a scale none of us predicted. In 1985, we had front-end scanners in 6 of our 14 stores. Today we're fully converted to scanners, and we're experimenting with a Direct Product Profitability system for space management and new product evaluations. DPP will require even more automation and more careful management and maintenance of all our computer operations.

In short, we've spent hundreds of thousands of dollars on new technologies, new service departments, and marketing programs to attract new customers. We're making big investments on the theory that sales volume will go up substantially—in fact, it *has* to go up for the investments to pay off. But all of these changes place new demands on our workers—and that's the big problem. They just don't seem able to live up to the new requirements.

Our customers consistently grade us high on quality, freshness, variety, and price. But we're getting creamed on service. Surveys show that our staff is slow, discourteous, and generally unfamiliar with our new products. The stores aren't clean enough and the shelves look jumbled. What good is a salad bar if half the containers are empty? Who can buy freshly baked bread if no one is behind the counter? Even Blue Hills Avenue and Sigourney Street, which have always been our best stores, are reporting more and more complaints.

Let me be direct here: Our business strategy isn't going to work unless we overhaul our personnel strategy. I don't have to tell you how tough it is to find and retain good people. Turnover exceeded 70% last year. We have a minicrisis with clerks and cashiers whenever one of the out-of-town people misses our morning bus—and someone misses the bus almost every morning. High school kids can't really do the job in the service departments, which is where we need people most, and the best ones seem to leave within a few months. Why bag groceries at Hilltop when you can get $7 an hour at McDonald's? And night stocking is a disaster. I walk through some of our stores first thing in the morning and they look like war zones.

The problems go beyond our union employees. I bumped into Ray Parento at the movies last week. I don't know about you, but for my money Ray has been one of our best store managers ever. He took over Sigourney Street

100 Bloomfield Avenue, Hartford, Connecticut 06105

DRAWING BY GARISON WEILAND

four years ago and he's done a fantastic job. Yet he told me that, for the first time in his life, he doesn't want to come to work in the morning. And when I think about our personnel problems, I can't say I blame him. I'm worried about Ray. If we start letting turnover in the ranks create turnover in our managers, we're in serious trouble.

Cutting labor costs isn't enough to keep us competitive. That's yesterday's news. The key for us now is to *raise productivity,* even if it means paying more to do it. We're probably going to have to pay more anyway. I just got copies of Local 413's new contracts with Homestyle Supermarkets and Pride's. The average raise is 13% over three years, and they even won increases in health and welfare benefits.

We need to take drastic action. All along, we've said that people look to Hilltop for jobs, not careers. Well, we've got to rethink that whole idea. We need to create career opportunities in the grocery business. After all, our last contract narrowed the wage and benefit gap so much between full-timers and part-timers there's no big cost advantage to hiring part-timers. What we really need is superior performance from our full-timers and more reliability from the part-timers.

If we went to Pete Ritter and offered to cut our part-timer percentage substantially over the next three years, my guess is that he would be willing to listen to new ideas for enhancing employee productivity. What we need in return is some flexibility on work rules and seniority. Nothing too extreme, but we've got to have more freedom to reward our best employees and create a sense of partnership among all our workers.

Here are three ideas I've been kicking around:

1. We should create a management trainee track for internal promotion of high-potential employees. Participants would spend six months in each of our service-intensive departments – produce, deli and bakery, fresh seafood – and then pick one to specialize in. Management trainees would be drawn from the bargaining unit; we don't want to hire outsiders. But the choice of people has got to belong to *us,* and we want to choose trainees based on merit – not seniority. We want candidates who demonstrate superior job performance and who display a long-term commitment to Hilltop.

2. We're always going to need part-time people, even if we agree to de-emphasize them. But we've got to attract better people and hold onto them longer, and that will mean some new programs. For example, Hilltop could provide tuition assistance for college students who work the 6 PM to 10 PM shift. We would pay monthly tuition stipends of $100 as long as the student continues to work for us. We could use a similar approach with high school juniors and seniors. Students who stay with Hilltop for at least one year and maintain a "B" average will receive a $1,000 tuition stipend. This lets us choose from the brightest students and it gives them an incentive to stay here. And it puts these kids' parents on our side.

3. I'm convinced the only way to boost productivity over the long term is to move away from hourly wage increases toward a pay-for-performance system. We've talked about the Super Fresh gainsharing program in Philadelphia. That's typical of how these plans work. Stores in financial trouble convince unions to accept wage cuts in return for bonuses for exceeding previously established productivity targets. (Since Philadelphia, Super Fresh has gone on to use gainsharing in Virginia and Maryland.)

GWIU

Grocery Workers
International Union
Local 413

February 22, 1989

To: Pete Ritter, President, Local 413
From: Jack Haskell, Shop Steward, Sigourney Street

At the last meeting of the local, you asked us to keep you posted on what's happening in the stores. It's about time the leadership did some communicating. Before you beat Tom DiSesa last fall, he'd run things for six years. In all that time, I don't think he visited Sigourney Street once. He sure never asked for any reports. This is just the kind of planning we need if we're going to deliver on your promise to win back what Local 413 lost last time—the best supermarket contracts in New England. Margaret Flynn won't know what hit her.

I want to fill you in on two things that happened last week. On Wednesday, Henriette Johnson got in a real brawl with some teenage kid who started about three weeks ago. You know Henriette—she's one of our most solid members, and she's been here 15 years. She works at her own pace and likes to do things her own way. She'd been on the express checkout for six hours, a couple of those hours without a bagger. This eager-beaver kid in the next aisle kept running at the mouth with the customers about school, boys, TV—anything that was on her mind. It was driving Henriette crazy. Finally, after she asked Henriette three different times for product codes she should have known, Henriette couldn't take it anymore. It got kind of ugly, and Henriette threatened to walk unless the kid was moved to a different register. Ray Parento wrote Henriette up, and she's still steaming. I told her I'd back her if she wants to grieve the thing, and I think we'd win, since we've won seven of nine grievances this year.

Then, I spent most of Friday trying to straighten out the meat department. A few months ago, the company decided to add a fresh-fish counter in the back of the store. Now Parento needs someone to run it. It's a hell of a lot of work (it takes 50 hours a week just to ice down and prepare the fish), not to mention that you smell like tuna all weekend. It turns out that Parento has been working on Pete Erwin, the new meat cutter who started about six months ago, and Pete was all gung ho about switching over to fish. But the new classification pays $50 a week more than meat cutting. We can't have a guy who's been here for six months making that kind of move. It took me all afternoon to convince our number-two meat guy to put in for the job.

Parento plays these kinds of games, then he wonders why our people complain that this place is screwed up. Does management ever ask *us* what's wrong? They open a big new deli counter and, instead of putting someone behind it, they set up a bell for people to ring for service. Then they won-

57 Marlboro Street
Windsor Locks
Connecticut
06096

der why the deli isn't doing the volume they expected! They whine about double coupons, and then they blow tens of thousands of dollars on an incompetent advertising campaign. *I* could have written better commercials. Hilltop ran three different specials in the last month where Pride's came in and offered deals on the same things at lower prices. It's pathetic.

There's no big secret to running this place better. If Hilltop wants to add departments, then they have to add people. If they want people to perform, then they have to pay them more. You know damn well Hilltop has the money. I can't believe what these yuppies will pay for some of this stuff. And why else are these Wall Street guys gobbling up supermarket chains left and right?

We all celebrated the new contracts at Homestyle and Pride's. I don't have to tell you how much support you have among the membership here at Sigourney Street. They're ready to go to the mat if that's what it takes to win.

I'm going to distribute surveys in a few weeks to get my people's views on the upcoming talks. But I don't need surveys to know what's on everyone's mind. We've got a stronger position than we've had in a lot of years, and it's time to stand tough. We should be able to make solid progress across the board.

Here are three things that should be at the top of the list:

1. We need a big wage bump. Full-time cashiers at Salvucci's are starting at $275 a week. That means they make just about what our people do. We really look bad on this one, Pete. Why should our members pay dues when nonunion cashiers earn the same money they do? I say we shoot for 18% over three years.

2. We've got to do better on health benefits. I hear the same things over and over again. People want lower deductibles on major medical, a raise in reimbursements on prescription drugs, and more comprehensive dental coverage. You guys at the local are the experts on benefit plans, but this has to mean a solid increase in Hilltop's contribution to the Health and Welfare Fund.

3. People want more paid time off. They're feeling burned out. The store is open two hours longer every day and coverage is getting worse—which means more pressure to work overtime. I've got two people who've put in 60 hours a week for the last three weeks. They like the money, but they also need some rest. Getting one more personal day and, say, birthdays would be a big hit with the people here.

I'll see you at the stewards' meeting next week. I've got a good feeling about this contract.

How Do You Bargain for Grade "A" Employees?

We asked participants in the supermarket industry and a labor-relations expert how Hilltop and its union might strike a productive bargain.

CAROL R. GOLDBERG *is president and COO of the Stop & Shop Companies and CEO of Bradlees Department Stores, both of which are based in Boston.*

Set the stage for progress next time.

In most service businesses, but particularly in a business as service-intensive as supermarkets, people are a company's most important asset. Sound like I'm endorsing apple pie and motherhood? Maybe so, but that doesn't make it any less true. The people challenge is the central *business* challenge facing Hilltop Stores. So it's reassuring that a planning memo on labor relations comes from Hilltop's senior executive for store operations. Human resource specialists can play vital support and implementation roles, but they cannot replace top management as the driving force behind employee policies.

The workplace issues that Ed Boyer has identified–rampant turnover, generational conflicts, lagging morale–may sound nightmarish to managers unfamiliar with the supermarket industry. But this is the gutsy reality that we struggle with every day. Conditions in many places are even *more* difficult than what Boyer describes. I know of supermarkets where turnover is double or triple Hilltop's 70% rate.

These conditions can take a very real toll on front-line managers. I recently had an encounter much like Ed Boyer's discussion at the movies with Ray Parento. A veteran store manager greeted me at the door of one of our stores and told me that his job just wasn't fun anymore. He had come up through the ranks in an era when labor was plentiful and training was on-the-job. If a new employee didn't work out, there were ten candidates lined up to fill the slot. This manager had been through a great deal with us over 30 years, but he was not prepared for the headaches of dealing with the changing work force. Soon after our discussion, he put in for retirement.

Ed Boyer is correct to emphasize employee productivity over short-term cost savings. A supermarket manager struggling in a tight labor market quickly learns to do whatever it takes to get bodies through the door. We regularly send recruiters to high schools and churches and have hiring tables in many of our stores so we can reach as many prospective candidates as possible.

The real challenge is retaining and motivating people once they come through the door. This is where much of Boyer's logic begins to break down. There is simply no way Hilltop can "cut its part-timer percentage substantially" over the next three years. Moreover, his argument for "careers" in the supermarket business may appeal to some of Hilltop's employees–but only to some. It is certainly not a basis for a companywide people strategy.

We think of our work force as composed of at least four distinct groups. There are full-time career employees of the sort Ed Boyer wants to develop, and we work hard to motivate and retain these people. There are also full-time employees who are attracted to the supermarket business at a particular point in their lives but who will never aspire to a career with us. There are part-time employees who will be with us for a limited period no matter what we do. Students are a classic example. Finally, there are "career" part-timers–mothers, retirees, and others who can't put in an eight-hour day but who expect to stay with the company for the long term. A human resource strategy built around the first group of employees will fail to meet the needs and priorities of these last three groups, who comprise the majority of a supermarket work force.

There is plenty of room for HRM innovation and creativity without trying to turn the supermarket business into something it is not. Hilltop must scrutinize every aspect of its operations and identify activities that could be performed more effectively on the outside. We increasingly think of our supermarket operations as more closely resembling the food *service* industry than the food *processing* industry. Stores need less highly skilled butchers and fewer of them; carcasses arrive preprocessed so that meats only have to be processed into final retail cuts. We don't wrap produce anymore, other than convenience items. Customers prefer bulk, and wrapping takes time.

So Hilltop can make great strides in productivity without radical changes in its labor agreement. That's important, because Ed Boyer and Margaret Flynn can't expect to win major changes this time around. He may have some terrific ideas (as well as some not-so-terrific ideas), but there's no time to evaluate or implement them. Hilltop has begun its labor-relations planning much too late. Do Boyer and Flynn understand the operations of Pride's and Homestyle thoroughly enough to know how a 13% wage increase affects their overall cost structures? Moreover, change on the scale Ed Boyer is proposing requires months, if not years, of informal dialogue

DRAWINGS BY CHUCK MORRIS

with the union leadership and rank-and-file. It certainly requires more than eight weeks.

Hilltop faces a second daunting problem. It's caught in the middle of a pattern bargaining cycle. Why should Local 413 agree to a contract dramatically different from what Hilltop's competitors signed—especially since, as Jack Haskell's memo indicates, the union thinks the company is quite profitable? Boyer and Flynn should make changing the contract's expiration date one of their top bargaining priorities. There is no need to sign a three-year agreement just because that's been the tradition. Hilltop should consider a one- or two-year deal that breaks the pattern and allows it to initiate more sweeping changes in the future.

Which is not to suggest there is nothing Hilltop can achieve in this round of talks. Tight labor markets do increase union bargaining power. But they have a more subtle effect as well, an effect to which Haskell alludes. When the labor market on its own provides wages and benefits comparable to union scale, unions have a more difficult time positioning themselves as third parties in the workplace. Today a supermarket union that doesn't work closely with management can get in serious trouble with its members.

We have found organized labor willing to negotiate and compromise to enhance store performance. No union will agree (nor should it) to Boyer's proposal for the near-abolition of seniority as a factor in promotion. But merit is important as well, and here is where Hilltop can win more flexibility in assigning, rewarding, and motivating employees. Stop & Shop's most recent contract gives us more flexibility to adjust wages to local labor-market conditions. We needed the freedom to accommodate those differences, and the union recognized that. With enough lead time, and if it can remove itself from the local bargaining pattern, Hilltop should be able to fashion the workplace innovations it needs to stay competitive.

GERARD L. GOOD *is president and CEO of Blue Cross and Blue Shield of Virginia, which is based in Richmond. In 1982, he founded Super Fresh Food Markets, a Philadelphia-based subsidiary of the Great Atlantic & Pacific Tea Company, and served as president until 1984.*

Don't confuse service with productivity.

Hilltop is in a bind. It has poured lots of money into its stores but has yet to see the increased volume it needs for the investments to pay off. There's no great mystery as to why. To a supermarket customer, the half-life of a good impression is about an hour. The half-life of a bad impression may be a year. When you lose customers because of bad service, they have to hear a lot of good things from a lot of different people before they'll give you a second chance.

That's why I'm so troubled by much of Ed Boyer's memo. He seems to believe that improving employee productivity will improve customer service. In fact, precisely the reverse is true. Good service means fewer problems in the stores, fewer disenchanted customers, and increased volume—which translates directly into more revenues per employee, the basic measure of productivity. Employee productivity at Super Fresh was well below industry norms in 1982, and it stayed there until volume grew because of our commitment to superior service. Had we aimed for productivity in the early stages, service would have suffered and volume would have stayed flat.

Boyer also seems to believe that shifting Hilltop's work force toward full-timers will somehow lead to superior employee performance. But pay classifications have no direct correlation with how effectively people work. A store demonstrates superior performance when the entire employee team understands the service objectives and develops programs to meet those objectives. Whether the team is composed of full-timers or part-timers will not much affect the end result.

Ed Boyer seems to think that Hilltop has serious labor problems. He's wrong. It has serious *management* problems. Hilltop doesn't need to make big investments in recruiting and training to improve service and build volume. Instead, management has to demonstrate the traits necessary for success in a service business today: openness, flexibility, and a willingness to adapt policies and procedures in response to employee input.

Senior executives in a service business must be open with middle management, hourly workers, and the union, if there is one. Openness implies a willingness to discuss future strategic and operating plans, current financial results, and all other issues of concern and relevance to the company's various constituencies.

Management often complains that organized labor doesn't understand the economics of its business, and these complaints are usually valid. But how can management expect unions to understand a business unless it is willing to educate them? This is just what we did at Super Fresh. We convened monthly meetings with representatives from senior management, the hourly work force, and the union. At the meetings, we reviewed financial results from each of our five geographic zones, asked people from each zone to discuss their stores' performances, and we described upcoming merchandising programs.

Wait a minute, you're thinking. Didn't the union use the financial information against us in the bargaining? Weren't we worried that the competition would get hold of our performance figures? The answer to both questions is "no." The Super Fresh gainsharing program created a real sense of mutual financial interest between management and labor—employees at each store benefited directly in that store's success. And we took some simple safeguards to guarantee confidentiality. For example, we presented financial results on overhead slides that meeting participants could review but could not take from the room.

Along with openness, management must show a willingness to change policies and procedures in response to employee suggestions. It may sound shocking, but your front-line workers, the people who interact with customers every day, really do have effective solutions to some of the thorny problems that we captains of industry try to solve from behind our desks. Of course, management also needs to develop a degree of tolerance and understanding. Certain employees or union officials will use participation to create trouble for management. But to reach the pot of gold, you have to be willing to tolerate the storm that creates the rainbow. Sometimes that's unpleasant. The end result is almost always worth it.

Recently at Blue Cross and Blue Shield, we decided to tackle some tough problems with employee training that had vexed us for some time. We assembled a task force of six people drawn from middle management and rank-and-file employees (we specifically excluded the top three management layers) and empowered them to talk with anyone, anywhere, about anything to do with employee training. The results were phenomenal. It was as if lights in our building that had been turned off for five years were suddenly turned on. People came out of the woodwork with ideas. Were half the ideas unworkable? Sure. But half of them were very constructive, and we changed our whole approach to training as a result.

Openness and participation do not mean that management forfeits its power to run the company. Service businesses must be demanding as well as solicitous of their people. If management communicates its performance expectations clearly to workers and demands adherence to these standards within a reasonable time period, people will rise to the occasion. Management may even find that many of its workers have higher and more demanding standards than its own.

So there are no quick or easy answers to Hilltop's dilemma. It's probably too late to make big strides toward openness and participation in the upcoming contract. But the company can use the change of leadership in Local 413 to make a fresh start and to lay the groundwork for more sweeping reforms in the next contract.

As the first step, Ed Boyer and Margaret Flynn should invite Pete Ritter and a few of his lieutenants to an unstructured and informal meeting. They should begin sharing information on finances and store performance, and they should ask for the union's help and advice on specific operating problems. The going may be rocky at first, and the union leadership may be distrustful, but Hilltop must begin creating a new climate of trust and openness throughout the organization. If it wants to be successful over the long term, it has no choice.

Workers will help—if management respects their interests.

JAMES T. BOYLE *is business representative of Local 464A, United Food and Commercial Workers International Union, which covers northern New Jersey and southern New York. He is a graduate of the Harvard Trade Union Program.*

If Ray Parento is one of Hilltop's best managers ever, then Hilltop is in even bigger trouble than Ed Boyer thinks. At most of the companies I know, conditions like the ones described in the memos would result in Parento's dismissal. These are challenging times for the supermarket industry, especially in the Northeast. Worker shortages, price wars, and Wall Street raids have put a lot of pressure on everyone. But incompetent management is the only way to explain many of Hilltop's problems.

To function efficiently, supermarkets must be properly staffed. Poorly maintained salad bars or bakeries without clerks defeat the very purpose of having them. That's a foolish way to save costs. If a minicrisis erupts whenever a cashier or clerk misses the bus, obviously there are not enough workers to start with. If stores look like "war zones" in the morning, where is the leadership and direction from the night-crew chief? Margaret Flynn and Ed Boyer should look inward, at their own policies and practices, before they push for radical changes in the labor agreement.

Boyer is right to focus on attracting and retaining talented employees, but several of his proposals leave room for improvement. For a part-time employee working 20 hours per week, a tuition stipend of $1,000 per year amounts to about $1 an hour. Why not just provide a wage increase of $1 an hour and an additional premium for the 6 P.M. to 10 P.M. shift, if that shift has the most serious retention problems? High school and college students are not the only candidates for these jobs, and increasing wages rather than offering stipends widens the recruiting range.

Boyer's proposal for internal promotion of high-potential employees also needs work. Training people for six months in each department means it will take two years for them to work their way through produce, deli, bakery, and seafood. That's much too long. Why not rotate them through a department every six *weeks*, and then let them make their choices? Also, eliminating seniority as a selection factor is a real nonstarter. Seniority is much more stabilizing than it is restrictive. It eliminates favoritism, both real and imagined, rewards workers for length of service, and provides a light at the end of the job-advancement tunnel. Seniority need not be the *only* criterion for participating in the management-training program, but Ed Boyer is much too cavalier about dismissing it.

As they develop their bargaining proposals, Margaret Flynn and Ed Boyer have to understand what the new realities in the supermarket industry mean for organized labor. The increasing diversity of the work force affects unions as well as management. Supermarket unions must advance the interests of a wide range of people, from teenage students who may be with a company for six months to 60-year-old heads of households with 30 years of service. The most direct way for a union to meet all its members' needs is to focus on the bread-and-butter issues–wages, benefits, working conditions, pensions, holidays–uppermost in everyone's mind.

Jack Haskell is right. Local 413 *is* in a strong bargaining position. It will not be disposed, nor should it be, to accept a host of experimental personnel policies and trade wage increases for gainsharing. What's good for one segment of the membership may not be good for another. The most effective bargaining strategy is to focus on gains that are shared widely across the work force. If management seriously expects to rewrite the contract, its proposals must reflect Local 413's difficult mission.

That said, organized labor is not oblivious to the serious challenges facing the industry. Our members are enthusiastic about new service departments like the kind Hilltop is introducing. They see them as an opportunity to upgrade their skills, and most employees go out of their way to make them a success. Moreover, these departments create employment. Technology has eliminated entire categories of supermarket jobs. In the past, for example, stores would receive whole carcasses that its meat cutters would prepare for sale. Today some meats are received portioned-controlled direct from the distributor. What has saved jobs that might have been eliminated and what has helped maintain our strength as a union are the new departments with which Hilltop is having such trouble. Workers will be willing to help management so long as management respects employee needs and interests.

It's time for a new spirit of cooperation.

DAVID LEWIN *is a professor of business and director of the Industrial Relations Research Center at the Columbia University Graduate School of Business. This year he is a visiting professor at UCLA.*

Hilltop is well behind the times in integrating its labor-relations and personnel strategies with its overall business strategy. Several real-world supermarket chains–including Kroger, Pathmark, Von's, and Safeway–have redesigned their compensation systems and introduced employee-motivation and productivity-enhancement programs to improve customer service specifically and business performance generally. Hilltop has much to learn from these examples.

Consider the issue of compensation. Ed Boyer wants Hilltop to implement gainsharing, a program with some real limitations. Gainsharing pays off for intermediate performance goals like improved store sales, lower costs, or increased productivity. In a fiercely competitive, low-margin industry like supermarkets, performance improvements induced by gainsharing don't automatically translate into improved *profitability*. Many supermarket chains prefer profit-sharing programs, since they pay off when total organizational financial performance improves. Ed Boyer and Margaret Flynn should consider profit sharing over gainsharing.

Of course, any incentive plan requires negotiation with Local 413. Jack Haskell's memo suggests that Hilltop's employees would be receptive to a profit-sharing initiative. Sigourney Street employees want to make more money and be more involved in store-level decisions. Assuming that Pete Ritter and other union leaders feel the same way–and the new president may be looking to deliver something new to his members–there may be a real opportunity for success here.

So Hilltop should explore the possibility of introducing a profit-sharing plan *and* an employee-participation plan that encourages worker involvement in store decisions. But senior management should keep two things in mind. First, Ed Boyer and Margaret Flynn must actively consult front-line managers like Ray Parento about the merits and consequences of incentive and participation plans. Store managers must be closely identified with the plans–they must "own" them and be their champions–or they could face challenges to their authority once the plans are implemented.

Second, the process of developing these plans creates real opportunities for labor-management cooperation. Jack Haskell is about to distribute opinion surveys to his members. Perhaps Hilltop management and Local 413 could *jointly* survey employees. Among other issues, the survey could explore attitudes on profit sharing and participation initiatives. If there is strong opposition, the plans need not be pursued further. But if there is support among the employees–and Jack Haskell's memo implies there might be–documenting worker attitudes could greatly enhance the chances of bargaining success.

Let me comment briefly on two other proposals. It makes sense for Hilltop to increase the ratio of full-time to part-time workers and to introduce career ladders of the sort Boyer is proposing. In return, Hilltop can insist on more flexible work rules and fewer job classifications. The company will end up with a somewhat older, less turnover-prone, and better paid (and thus more costly) work force. But such a work force is also more likely to display the motivation and generate the performance improvements that Hilltop seeks. Creating less dependence on younger, entry-level employees also makes good demographic sense. The 15- to 24-year-old age group is the fastest declining group in the United States, and shortages of entry-level workers are likely to become very pronounced by the early 1990s.

Ed Boyer also recommends that Hilltop push for storewide seniority in place of companywide seniority. This is a mistake. He's worried that older workers will bump younger workers from stores where gainsharing yields big bonuses. A companywide profit-sharing plan mitigates this concern. And companywide seniority may work to Hilltop's benefit. More experienced employees can undertake new departmental or promotional assignments, regardless of the store where openings arise, which is consistent with the concept of career ladders.

Finally, whatever it agrees to at the bargaining table, Hilltop should begin a new employee-orientation program emphasizing that the company is in the customer-service business, not the food business, and that employees, not technology or prices, are what make the competitive difference. I would accompany this orientation program with a recognition program (such as employee-of-the-month) that features a cash or in-kind award for outstanding performance. This will help solidify employee commitment to high-quality customer service.

Reprint 89212